PRESENTING SCIENCE

# Presenting Science
## A practical guide to giving a good talk

*Çiğdem İşsever and Ken Peach*

OXFORD
UNIVERSITY PRESS

# OXFORD

UNIVERSITY PRESS

Great Clarendon Street, Oxford OX2 6DP

Oxford University Press is a department of the University of Oxford.
It furthers the University's objective of excellence in research, scholarship,
and education by publishing worldwide in

Oxford   New York

Auckland   Cape Town   Dar es Salaam   Hong Kong   Karachi
Kuala Lumpur   Madrid   Melbourne   Mexico City   Nairobi
New Delhi   Shanghai   Taipei   Toronto

With offices in

Argentina   Austria   Brazil   Chile   Czech Republic   France   Greece
Guatemala   Hungary   Italy   Japan   Poland   Portugal   Singapore
South Korea   Switzerland   Thailand   Turkey   Ukraine   Vietnam

Oxford is a registered trade mark of Oxford University Press
in the UK and in certain other countries

Published in the United States
by Oxford University Press Inc., New York

© Çiğdem İşsever and Ken Peach 2010

The moral rights of the authors have been asserted
Database right Oxford University Press (maker)

First published 2010

British Library Cataloguing in Publication Data
Data available

Library of Congress Cataloging in Publication Data
Data available

Printed in Great Britain
on acid-free paper by
CPI Antony Rowe, Chippenham, Wiltshire

ISBN 978–0–19–954908–5 (Hbk.)
978–0–19–954909–2 (Pbk.)

1 3 5 7 9 10 8 6 4 2

*To our families and our friends*

# Preface

This book started out as a course given to the first year graduate students in particle and accelerator physics in the Department of Physics at the University of Oxford, where we tried to pass on some of our ideas about how to present science in an interesting and entertaining way, without sacrificing rigour. The students were asked to prepare 10-minute presentations on a subject of their choice, which were rehearsed with one of us and half the class, and feedback given. The revised presentation was then given to the whole class in a small conference-style session. Many of the initial presentations were excellent—indeed, we have ourselves learned from our students. But in a few cases there was the famous 'quantum jump' between the rehearsal and the final presentation. We would like to thank all of the students who took part in this exercise.

Most of the examples in this book come from particle physics—this is what we do professionally, and it is easy for us to find examples (good and bad) from our own talks and those of our colleagues. (The bad examples have been suitably anonymized.) However, we do not believe that the principles governing presentation of science depend upon the particular area of science being presented. We have, in a number of cases, added a footnote to help the non-expert understand the science underpinning the example, but in general we have left it at the level of 'pretty pictures'.

We would like to thank Steve Lee from the Physics Department's Media Services Unit for taking the photograph for the cover, Rudi Apolle, Chris Boddy, Caterina Doglioni, Penny Jackson, Sarah Livermore, Alex Pinder, Sophie Redford, Suzie Sheehy and Sam Whitehead for being a willing audience, Maria Fiscaris for help with the figures, and Cecilia Uribe-Estrada for help locating the Cervantes quotation. We would like to thank April Warman and Sonke Adlung from the Oxford University Press for their advice and forbearance. We would like to thank colleagues in Oxford for allowing us to give the course on presenting science; it has 'informed, educated and entertained' us enormously.

Most of the figures used to illustrate the text have been specifically created for the book, but we are happy to acknowledge the following sources. Some of the images in Figs 5.2, 5.12, 5.13 and 5.15 come from 'ATLAS Experiment at CERN', http://atlas.ch and CERN. The background image in Fig. 5.9 comes from STFC. The tables in Figs 5.28 and 5.29 come from the ATLAS Technical Proposal, CERN/LHCC/94-43. The data plots in Figs 5.25 and 5.30–5.33 are standard plots illustrating the scientific reach of the Large Hadron Collider at CERN. Figures 3.4 and 5.9 use images from Microsoft's Media Elements.

Finally, we would like to thank you for reading this book; we hope that it will enable you to develop your presentational style, and help you to give 'a good talk' every time.

Çiğdem İşsever
Ken Peach
Oxford, August 2009

# Contents

1   **Introduction**    1

2   **Preparatory work**    4
2.1  The audience    4
2.2  The structure of the event    5
2.3  The venue    5
2.4  Laptop, memory stick, plastic foils or chalk and talk?    6
2.5  What are the slides for?    7
2.6  Scoping the talk    9
2.7  The process of writing the talk    10
2.8  Summary    12

3   **Structure of the presentation**    13
3.1  The structure of the content    14
3.2  The backup slides    24
3.3  Summary    24

4   **Identifying the context of the presentation**    25
4.1  Some common contexts    26
4.2  Assessing the audience    38
4.3  Judging the length of the presentation    40
4.4  Timing    41
4.5  Am I the first speaker, or the last speaker, or somewhere in the middle?    42
4.6  Summary    43

5   **Style**    44
5.1  Font    44
5.2  The frame    47
5.3  Bullets and boxes    58
5.4  Pictures, plots and tables    64
5.5  Animation    68
5.6  Equations    71
5.7  Posters    77
5.8  Visual aids and props    79
5.9  Summary    80

6   **Preparation and presentation**    82
6.1  Position to audience    84
6.2  Voice and language    86
6.3  Microphones    86

| | | |
|---|---|---:|
| 6.4 | Using the laser pointer | 88 |
| 6.5 | Dress code | 90 |
| 6.6 | Entertaining the audience—jokes and asides | 90 |
| 6.7 | Are you sitting comfortably? Dealing with nervousness | 93 |
| 6.8 | Rehearsal | 95 |
| 6.9 | Dealing with questions | 97 |
| 6.10 | Final remarks | 101 |

**7   Concluding remarks**   102

**Appendix A   Presenting complicated equations—a worked example**   104

**Appendix B   Some PowerPoint tips**   111
| | | |
|---|---|---:|
| B.1 | Blanking out the screen | 111 |
| B.2 | Running out of time | 111 |
| B.3 | Saving space with pictures | 112 |
| B.4 | PowerPoint and equations | 112 |

**Appendix C   Meeting the media**   113
| | | |
|---|---|---:|
| C.1 | After a press release | 114 |
| C.2 | The unsolicited contact | 115 |
| C.3 | A final word on the media | 116 |

**Index**   117

# 1
# Introduction

*To inform, educate and entertain*
*From the mission of the BBC*

'Giving a talk' is part of everyday life for the working scientist—it is one of the most important ways in which we communicate our research and our ideas to others. The 'talk', of course, covers an enormously wide range, from a 10-minute briefing on progress to a handful of colleagues at the regular Monday afternoon group meeting to a keynote address to a major international conference with more than a thousand delegates. However, whatever the occasion, the aim of the talk is the same—**to get your message across to your audience clearly and effectively**. Members of an audience usually appreciate a good talk. Often, someone will come up afterwards and say something like 'Thank you very much for that talk; I really enjoyed it and I learnt a lot.' When that happens, you have achieved at least some of your goals; you have informed, you have educated and you have entertained.

Presentational skills are becoming more important in all walks of life. Presenting science, professionally to other scientists and to a wider public, has particular issues that need to be addressed. Our aim in this short book is to equip you, the reader (undergraduate, postgraduate, professional researcher, junior academic . . . ) with the basic skills needed to make a good presentation. Our approach is essentially pragmatic and not dogmatic, with emphasis on four essential points:

1. The goal is to **communicate** the science to the audience.
2. The **speaker is responsible** for everything that appears, and does *not* appear, on each slide.
3. The **structure** and **appearance** of the presentation, as well as the **content**, are part of the communication process.
4. There is no standard way of doing things—each slide, and each presentation, is different, and must be tailored to the **needs of the audience**.

Before you read further, we need to make one thing clear—we do not pretend to be able to give you 'Five Simple Steps to the Perfect Presentation', and we are somewhat sceptical about any such approach. There are several reasons why we take this stance.

- It is *your* presentation, and it must be coherent with *your* personality and reflect *your* style. This may seem to be an odd point to make, but it is crucial to our approach. If the style of your presentation is at odds with your personality, then people are likely to suspect that it is not your talk, and have a tendency to discount, perhaps unconsciously, your conclusions.

- While we have our views about what makes a good slide, and what makes a poor slide, when we try to transform these into rules (e.g. use 24 point bold Arial yellow letters on a blue background for the slide title) we can always think of circumstances where this might be just the *wrong* thing to do to achieve the impact that you desire.

- We scientists are a sceptical bunch ... systematic doubt is at the heart of the scientific method. We therefore tend to distrust a presentation that is *too* perfect. We like to think that we are getting the latest information, hot from the microscope, and are prepared to tolerate the odd rough edge to the presentation. Of course, this does not mean that we scientists enjoy a *poor* presentation, but that we should not invest a great deal of effort once it is good enough.

So, use this book to develop *your* style. This is something that is, or should be, unique to you, and will say something about you, your personality, your subject and your competence. As you develop your style, take time to sit at the back of the hall and look at the slides—does my style help the audience understand the subject, or does my style get in the way of the message?

A poor presentation of a fascinating subject may well fail to convince. An exciting presentation of some rather pedestrian information may be very well received. This is not to say that style is more important than, or a substitute for, substance, but it is to emphasize that you owe it to your audience (who usually do not have to sit there) to make it is as easy as possible for them to understand your message.

The book will assume basic familiarity with computers and some familiarity with a presentation manager such as Microsoft$^{\textregistered}$ Office PowerPoint$^{\textregistered}$. The examples in this book will use PowerPoint.

## Structure of this book

In the remainder of this book, we will take you through the various stages in the preparation of a good talk. These chapters can be read in almost any order. Chapter 2 discusses the preparatory work that you need to do before sitting down to write the talk. Chapter 3 discusses the basic structure of a talk, and Chapter 4 discusses the context within which the talk is to be set. Of course, the one influences the other, and so these chapters should probably be read together. Chapter 5 discusses the style of the presentation (font, border and background, pictures, plots and diagrams, animation, etc.). Chapter 6 offers advice on the preparation for the presentation, and in the last chapter we summarize the process of writing a good talk, and draw some conclusions. The Appendices contain a few technical details and tricks that we have picked up along the way, which might be of more general interest.

Throughout the book, we have highlighted in bold certain words or phrases that we think represent key ideas or important issues. As you read the book, you might wish to make a list of the concepts that you think are important, and which will help you develop your style.

Finally, 'giving a talk' is stressful, even if you have given the talk many times before. In fact, if you ever lose the stress, perhaps you have also lost the enthusiasm that should accompany any talk. The audience should know that you *care*. It is rumoured[1] that Enoch Powell, the rather controversial Conservative politician, noted orator and classicist, believed that politicians should speak with their bladders half full, to give a sense of urgency to their speeches—we do not recommend this approach.

---

[1]For example, it is quoted in the Times Online, September 16 2005 http://www.timesonline.co.uk/article/0,,3-1782828,00.html.

# 2
# Preparatory work

*Estar preparado es la mitad de la victoria*[1]
*Miguel de Cervantes (1547–1616), Don Quijote de la Mancha*

It is impossible to overstate the importance of good preparation for a talk. Even when the subject is very familiar, the **preparation time** is likely to be at least ten times the length of the talk, and if it is a major presentation, you should be ready to spend even longer. It is only with the third or fourth time that you give the same talk (or effectively the same talk) that the preparation time *might* take less time than giving the talk.

## 2.1 The audience

It is important to know your audience. Even if some of the slides are the same, the talk will be very different depending upon whether it is given to your professional colleagues (for example in a colloquium or seminar[2]), to a general audience (for example a lecture to the British Association for the Advancement of Science) or to a group of school pupils. You will need to know how big the audience is likely to be, whether there are any special guests (should you begin 'Your Majesty'), and if there is any special connection between the audience or institution and the topic of your talk (was your host an author on the key 'discovery' paper).

What do you expect the audience to know about the subject in advance? In general, the more 'public' the audience, the less you should assume they know. However, beware of assuming that the professional audience knows a great deal about your special subject. One useful trick

---

[1] To be prepared is half the victory.

[2] A colloquium may be defined as an academic meeting led by a different lecturer and on a different topic at each meeting, or a gathering of scholars to discuss a given topic over a period of a few hours to a few days. A seminar may be defined as a small-group teaching situation in which a subject is discussed, in depth, by the participants. In practice, there is little difference between the two.

for **colloquia** in university departments is to say to the organizer or Head of Department something like 'I have prepared a few introductory slides for the benefit of the graduate students—do you think that I should show them?' The response is usually something like 'Ah yes, the graduate students—I think that it might be a good idea to show them!'

As well as knowing who your audience is, it is necessary to know *why* they have invited you to give the talk, or *why* they have taken the time and trouble to come along to listen to you. Until you know this, it is difficult to see how you can set about meeting their expectations. In general, the audience is there voluntarily and, if they do not like what they hear they will become restive and *in extremis* will leave. We have attended talks at major conferences where the half-life of the audience was about 10 minutes!

How big is the audience likely to be? Under ten? Around thirty? More than 100? More than 1000? This affects the way you address the audience—it is difficult to be simultaneously intimate with 1000 people, and half a dozen may not appreciate being addressed as a political rally. (See also section 4.2).

## 2.2 The structure of the event

What will the structure of the event be? Will there be a formal introduction and a vote of thanks at the end? Will questions be invited from the audience? How critical is the timing? Does '1 hour' mean starting 5 minutes past the hour and finishing 5 minutes before the hour, with 5 minutes of questions, so that the real **length of the presentation** is 45 minutes and not 60 minutes? Or do they want the full hour, with 30 minutes of informal questions over tea and biscuits afterwards? Again, you need to know this before you prepare yourself to give the talk, even if it is a talk that you have given several times before.

Does the invitation include lunch or dinner? If so, who will be in the party? Does the meal come before or after the talk? Particularly if the meal *follows* the talk, you may need to have some more details about the topic available beyond what was in the presentation. You also need to know if there is some ulterior motive for the meal, especially if it is dinner—are they thinking of offering you a job?

## 2.3 The venue

You need to know, well in advance, the type of venue and audio-visual equipment that is available. Most venues these days will have a modern data projector, but may not have a wide range of accessories. If you have special requirements (for example, if you have a Mac you need an adapter cable for most data projectors), check beforehand or make sure that you have one with you.

Of course, on most occasions you will know the answers to these questions without having to ask, but they will, or should, influence the way you prepare the talk. If you do not know the answers to any of these questions, make enquiries—your hosts will appreciate your professionalism, and you may, one day, avoid serious embarrassment.

## 2.4  Laptop, memory stick, plastic foils or chalk and talk?

Although the focus of this book assumes that you are preparing a computer presentation, there will no doubt be occasions when other media are more appropriate, and we need to consider the alternatives. The first point to make, of course, is that much of the preparation does not depend critically upon the **medium** of its delivery. Even if you give a 'chalk and talk' presentation, it is still useful to have an idea of what the blackboard should look like as the presentation evolves.

The choice of presentation medium depends partly upon the material that you wish to present, partly on your style and personality and partly on what your hosts expect. Talks given from the laptop or computer are usually able to pack in much more material than with foils or slides, and writing on a blackboard or whiteboard can be slow. However, if the material is very dense (for example, a mathematical derivation of some complexity), then there may be advantages in using the older technology. Even so, with proper self-control, and judicious use of animation, it should be possible to present the material equally effectively from the computer.

These days, there are many occasions where a talk has to be given via a video conference or over the telephone. In such circumstances, it is essential that the slides are clearly numbered and labelled (so that you can say 'we should now be on slide 9'), and animation generally should be avoided.

If you are giving a computer presentation, there are several options— using your own laptop, loading your talk onto the local computer from a memory stick or CD, or from a web-site agenda page. For smaller venues, for example a departmental seminar, it is usually possible to use your own laptop (although your hosts may like to have a copy of your talk for their archives); this means that you can be reasonably certain that the slides will appear on the screen very much as you expect. For larger venues and major conferences, the organizers these days are likely to insist that you use the venue computer system—it is just too disruptive to waste 5 minutes at the beginning of each talk synchronizing the laptop with the data projector (and sometimes failing). This *usually* works, but you should take some precautions—do not use too much fancy animation or embedded video-clips, and be sure to embed the fonts with OPTIONS/SAVE/EMBED TRUETYPE FONTS. Even so, different versions of the software can lead to some surprising effects. Using pdf files is more likely to avoid most problems, but this may lose some of the special features of your favourite presentation manager that you like.

It is very important, if you need to use an overhead projector or chalk, that you check with your hosts beforehand. In many places these days, the overhead projector lies abandoned in the corner, dusty and scratched and with no spare bulb, used mainly as a coffee table.

**Mixing media** can, if well done, be very effective—the main part of the presentation from the computer, but with particular points amplified on the blackboard. However, this is quite difficult to do well, especially in an unfamiliar auditorium. These days, the lighting is usually set for a computer presentation, and the blackboard poorly lit. While there are often 'blackboard lights' to improve visibility on the blackboard, these then render the computer images less readable, and switching them on and off is a distraction.

As with everything else, if you think that it helps to convey the message, then do it. But perhaps consider carefully whether the message can be conveyed just as effectively with less fuss.

## 2.5 What are the slides for?

There are essentially two extreme views about the purpose of each slide:

1. It is there to make sure that the speaker does not forget any important point.
2. It is there to help the audience understand the subject.

Now, the second view *should* be the main motivation for each slide, but in practice it seems as if the first is quite commonly considered almost as important. There is always a temptation towards the former when the speaker is nervous for one reason or another. The working language for many international conferences is English, and 'giving a talk' presents special challenges for speakers whose mother tongue is not English. If English is your mother tongue, you should also consider the needs of those in the audience for whom English is a foreign language. It so happens that the way to address both of these issues is the same—use simple words, avoid long strings of words, make sure that the font is large enough to be clearly visible from the back of the hall, and avoid confusing backgrounds. We will discuss the need to speak clearly in section 6.2.

Whatever the occasion, just reading from the slide is rarely a good idea. The problem is that the audience does not really know what to do—should they read the slide for themselves and forget what the speaker is saying, and perhaps miss the important aside? Or should they ignore the slide and just listen to the speaker, and perhaps miss the important diagram or figure?

Presentations are very often these days placed on a web-site for those who attended to consult later, or for the benefit of those who could not attend. Without the speaker's words and guidance, the presentation may not make much sense. However, designing the talk to be understood 'standalone' risks favouring those who could not be bothered to turn up over those who took the time, and paid you the compliment, of listening to the presentation. There are at least two better ways to make the presentation accessible to those who are interested in your topic but who were, presumably with good reason, unable to attend.

1. Produce a **written set of notes** for each slide—this is relatively easy in PowerPoint using the Notes feature, but a simple text file can serve if this feature is not available in your preferred presentation manager. Note that we would *not* recommend writing out the text of your speech verbatim, but rather point out the important details that the slide illustrates. Of course, these notes will be invaluable for you in preparing the talk, especially if you are nervous. It is important that these are genuinely 'notes'—if you do lose your way, you need very quickly to find the appropriate point in the presentation, and this is almost impossible if it is densely written text.

2. Provide a link on the title page to a written paper (or to several papers) covering the material of the talk.

**Remember always, the purpose of the slides is to help the audience understand the subject**. Once you start to relax on this and make the slides serve some other purposes (like being intelligible to those who were not there) you risk confusing the audience.

While discussing the purpose of the slides, let us address another issue—the **corporate style**. The purpose of the corporate style is to sell the corporation. This poses a problem for the conscientious communicator, if his or her superiors insist on the corporate style. Ideally, a slide should carry a single message or theme; if it carries more, the audience tends to become confused—what is this slide trying to tell me? If the purpose of your slide is to present your key result, you do not want this message obscured by what is in effect advertising for your employer. We will return to this (see section 5.2.3) when we discuss the slide 'frame' in a later chapter.

In preparing the slides for your talk, you should keep clearly in your mind your **responsibilities to the audience**. These responsibilities can be simply stated.

1. You are responsible for everything that appears on the slide.
2. You are responsible for everything that *could* have appeared on the slide but that does *not*.

So, the basic message of this book is that whatever appears on the screen (font, size, character, colour, placement, pictures, graphs, animation, extraneous information) should in all cases be the result of a conscious choice, and not because it is the system default, or because you borrowed a slide from a friend, or because you picked one from the web. Even if one of these is the basis for your slide, it is still worth editing it to be consistent with your style—in so doing you will also absorb the material and thus be better able to speak confidently when you reach this point of the talk.

## 2.6   Scoping the talk

Once you know who the audience is and why they might be interested in what you have to say on the topic (whether of your choosing or of theirs), you can scope the talk.

A good starting point is the **'Conclusions' slide**—why are you giving the talk? What is the message? This is, after all, the slide that will (we hope) end with enthusiastic applause. With this in mind, it is useful if all of the conclusions can be contained on one slide. This may not always be possible, but it should always be an aim.

Knowing your audience tells you where you start, and now you know where you finish. The next thing to do is to plan the route that gets you, and your audience, from the start to the finish. This 'route' is in effect the **'Contents' slide**. Having done this, you have the scope of the talk. The task of writing the talk itself can begin.

There is *always* more material that you *could* present than can be sensibly included in the time available, so you need to make a choice. Do you try to include as much material as possible, and hope that the audience, or at least some of them, can keep up (Scylla), or do you make sure that you have enough time to explain clearly all of the points, but skimming over some of the more tricky issues (Charybdis)? How you choose to steer between these two monsters depends upon the audience. But remember, if the audience does not have time to absorb the information on the slides, they are unlikely to gain much from the talk as a whole, even if they are all experts. In this situation, **backup slides** provide a solution. A set of well prepared backup slides can help you respond to expert criticism or to questions about certain aspects of your work which you had no time to discuss in detail (see also section 3.2).

Now you have all of the information that you need to choose the **title of the talk**, assuming that this has not been given to you by your hosts. The title should give the audience information about the subject, level and approach of the talk.

Now turn on the computer.

## 2.7 The process of writing the talk

Later in this book, we will help you with the technical details of constructing a good talk. In this section, we want to address the *process* of writing the talk. As with everything else, there are no real rules that we can give you. You have to develop your own system, one that you find works for you.

Once you have thought of a working title, decided upon a few tentative conclusions and mapped out a possible way to progress from one to the other in a reasonably logical way, you have to start actually writing the slides. There are many ways of doing this, all having advantages and disadvantages. Here are a few ways that you could try.

- You can start at the beginning, work through your route to the end, and then review and revise.
- You can start with the key slide and work forwards and backwards, reviewing and revising as you go along. In professional presentation courses, this slide is sometimes called the 'money slide', but this is usually inappropriate in a scientific context, unless of course it is a grant proposal, in which there is usually quite literally a 'money slide' that you had better get right.

- We suppose that you could also start at the end and work your way back to the beginning. We haven't tried it, but we have no reason to believe that it would not work ... at least you know where you are going!

- You could always take half a dozen slides from a few previous presentations on this or similar topics, throw them together in a vaguely logical order and hope that no-one has been to any of the previous talks. We are not actually advocating this method, but it is clearly one that is used by a significant number of people, judging by some of the talks we have sat through, and we may even have used it ourselves from time to time. (For example, when our boss pleads 'Help me ... I have half a dozen distinguished visitors this afternoon and I need to give them an introduction to "Instrumentation for Quantum Gastronomy"; can you let me have something by coffee?')

- There is perhaps a more respectable way of reducing the amount time it takes to prepare a new talk, and that is by having a library of talk fragments for standard segments. For example, it would be possible to have the history of the subject, the derivation of the basic formulae, a review of the current state of knowledge and description of the experimental set up as ready-made segments to be selected as appropriate 'chapters'. However, there are at least two disadvantages to this approach. Firstly, as you are asked to give more talks on your chosen subject, your audience may also become equally familiar with your approach, and so become less likely to spot the really new and exciting part of the presentation. Secondly, the presentation may seem somewhat disjointed, as if it had been assembled rather hastily from the material to hand (which is, of course, precisely what has happened).

- Finally, there is the method of starting from the last talk that you gave on the subject, updating a few of the slides and hoping that not too many people will have seen it before. This has the merit, at least from our observations, of being very commonly employed, although we cannot really recommend it.

Whatever way you choose, remember to keep in mind the purpose of the talk, which is to help the audience understand and appreciate the subject, and if possible in a way that gives them pleasure.

## 2.8 Summary

There are several things that you need to consider before sitting down to write the talk, which if done carefully will save time later. This includes understanding who will be in the audience, what type of event is involved, where it is taking place, how you intend to deliver the talk, the scope of the talk and how to start the process of actually writing the slides.

# 3
# Structure of the presentation

*La dernière chose qu'on trouve en faisant un ouvrage, est de savoir celle qu'il faut mettre la première*[1]
*Blaise Pascal (1623–1662), Pensées*

Just as a house consists of a harmonious arrangement of some common elements (door, hall, kitchen, lounge, bedroom, bathroom, garage), so a presentation consists of a set of standard components (title, outline, introduction, message, conclusion) arranged in such a way that the audience can be informed, educated and entertained. The **structure of the talk** is important in keeping the audience engaged—a poorly structured talk leaves the audience confused and disorientated.

And just as with architecture, there are some conventions. In a house, the main door is usually near the front of the house, the bathrooms tucked away at the side, the kitchen somewhere near the dining room, the lounge leading to a pleasant garden, the bedrooms upstairs and so on. Similarly, a talk usually starts with a **title slide**, follows with an **outline**, continues with the **introduction**, delivers the **main message** and reaches some **conclusions**. But sometimes, as with architecture, you may wish to depart, perhaps radically, from this conventional structure—for example, starting with the conclusions because you want the audience to *know* where you are taking them. However, as with architecture, you need to be sure that this departure from convention serves a purpose—to help the audience understand your message. Failure to structure the talk properly risks reducing your message to a pile of rubble—unattractive, unappreciated and soon forgotten.

If the structure of the talk as a whole is like the architecture of a building, the **structure of the slides** is like the interior decoration of the rooms. Each room, and each slide, is different, but all usually share some common features. In general, slides should have a title which tells the audience what the slide is about, just as, for example, it is often

[1]The last thing one knows in constructing a work is what to put first.

useful to label the bathroom as the bathroom—obvious but, if you are new to it, helpful. The body of the slide contains your message; a figure or figures, lists, tables, pictures, etc. We will discuss the design of the slides in Chapter 5. *We note here that the structure of the slides and the structure of the talk must be harmonious.*

While there are some conventional ways of giving a talk, we are cautious about being too prescriptive. Our only real concern is that whatever choices you make, they are *deliberate*, designed to create an effect that will help the audience. In general, as in this chapter, we will describe the conventional structure of a talk, one which, through observation and personal experience, we think works well under most circumstances. We will also point out some alternatives, and give examples of where an alternative approach might be acceptable or even advisable.

The details of the structural elements will depend on the length of the talk. We will now discuss these building blocks of a presentation in more detail.

## 3.1  The structure of the content

Before starting to write the slides, you need to take a step back. The most important step is to think about the structure of the talk; what do you want to put into it, what do you want to leave out, what needs to be introduced first in order to prepare the ground for the later topics. In general it is confusing for the audience, unless they are all experts, to speak about things without first defining what they are. This is especially true for conferences. If your audience is an expert audience you will not need to introduce too many things, but if your audience is a lay audience you will have to take their knowledge level into account otherwise you will fail to get your message across; they will remember that you were very enthusiastic about something, but will have only the vaguest idea what that something was.

A talk has in general most of the following sections:

1. an introduction to and motivation for the subject—this includes the title slide;
2. an outline of the presentation;
3. a description of the technical background to the talk;
4. a description of the method;
5. the presentation of the results or the key message;

6. a summary of the presentation and an outline of future work, and any conclusions or recommendations that might follow;

7. backup slides.

There is no rule which governs how long each of these sections should be. Returning to the 10-minute **weekly progress report**, perhaps only the title slide and the fifth and sixth items of the above list might be needed, although even here if you are presenting something for the first time to the group, elements of the third and fourth items might be needed as an *aide-memoire* to your colleagues, who may not have been following your work all that closely (until now, of course). For a **professional seminar**, you might wish to spend about half the talk presenting and discussing the methodology for obtaining the results, the results themselves and their significance, but for a more **popular lecture**, you will probably need to spend more than half the talk on the first three sections, or else the audience might find it hard to understand the significance of *your* contribution to the field.

*Transitions between different parts of your talk.*   Your talk should have a rhythm, where one topic flows gently into the next, unless of course you deliberately want to startle them out of their comfort zone for some dramatic purpose. However, in general you should not jump around too much. You should also plan for some natural **pauses**—your audience can only take so much information without being temporarily saturated. If the talk is longer than 10 or 15 minutes, try to break it up into 10- or 15-minute segments. It is good practice to follow a difficult or detailed section of your talk with a lighter passage which is easier for the audience to digest, and so absorb the information. This can be achieved in many ways. You could summarize what you just have presented as a kind of 'instant revision' to underline the key message. You could, if you are good with humour, tell a relevant **joke** or amusing anecdote, again to give the listener something to hold on to, to help him or her remember the point. If the previous section was not so difficult, you might still give them a little rest by giving a brief introduction to the next section of your talk. We will come back to this later, but it is helpful to the audience to indicate that this is a **short break** with the aid of some body language—if you have a radio microphone or no microphone at all, move away from the lectern, or lean backwards, or use some other method which distinguishes the core of the presentation from this short aside, and then, with an exaggerated gesture and perhaps a comment, 'get back to the talk'.

### 3.1.1 The title slide

The title slide sets the stage for your presentation. Its function is to catch the attention of the audience, engage them with the topic, introduce yourself and prepare the audience for what will be coming. In most cases, you should have the title of your presentation on this first slide, together with your name, your affiliation and your position, the date and the venue.

*The title.* Let us begin with the title. This should represent the **subject of your talk** and give the audience an idea of the approach or **style of the talk**. It is worth taking the time to choose a good title—you will have put a lot of effort into preparing the talk, and you might as well let as many people as possible benefit. The title can tell them not only what you are going to talk about, but also how you are going to present it. The experts will presumably come anyway, but you really would like to attract those who perhaps do not know as much about your subject as they should. The title will appear on the conference or seminar programme, and should make people think 'I would really like to go to that' instead of doing something else—we all have many things to do.

*An example.* Let us suppose that you have been asked, as the world's expert on Quantum Gastronomy, to give the keynote lecture at a major international conference on the subject. Consider the following titles— which would tempt you away from the bar?

1. Quantum Gastronomy
2. Quantum Gastronomy—an Overview
3. Quantum Gastronomy—Status and Prospects
4. Quantum Gastronomy—a personal perspective
5. Adventures in the Quantum Kitchen
6. Changing the Paradigm—Quantum Gastronomy at the Crossroad

The first title gives no clue, beyond the subject, while the second and third promise a comprehensive if slightly pedestrian review. The fourth title is a little more interesting, but might be idiosyncratic and could be provocative. The fifth promises entertainment, a lively introduction to the subject, and may be difficult to deliver. The last title hints at something potentially earth shattering and, again, you might find it hard to deliver.

So, choose your title with care, and match it to your purpose in giving the talk, and also to your personality—do not promise more than you can reasonably deliver. Of course, if the talk is the regular report on progress to the group, why bother with anything more than 'Progress Report'?

*Introducing yourself, the occasion and the date.*   After the title, you have a choice, and which you choose will depend upon the context. Usually, you will wish to introduce yourself and perhaps say where you come from (who pays your salary), and probably follow this with the occasion and the date. However, for a particularly prestigious address you might want to identify the occasion first. If you give the talk again, *remember to change the occasion and the date*—audiences like to think that you have prepared the talk especially for them. You also need to decide how to introduce yourself—this will tell the audience something about how you see yourself, and how you expect them to see you. For example, with an increasing degree of formality, you might describe yourself as *Joe Bloggs, Dr J. Bloggs, Prof. Joe Bloggs, Professor J. M. Bloggs* or *Professor Joseph M. Bloggs, FRS.* Which you decide to use will depend upon you and your personality, and on the occasion (the context—see chapter 4).

*Other information.*   Other information that you might wish to include on the title slide includes a contact e-mail address, or a web-site address where the talk can be downloaded or seen again, or where further information might be obtained.

A simple title slide is shown in Fig. 3.1.

What else you choose to put on the title slide depends again on the context of the talk. In these image-conscious days, it is quite usual to see a seasoning of logos (university, department, group, collaboration and, if you are so inclined and entitled, an armorial coat of arms) and perhaps an iconic photograph or two. How much of this you include is up to you, but remember that the function of the title slide is to prepare the audience for what follows ... if the garnish helps, then do it; if it is more likely to intrude, think twice.

*Using the title slide to start preparing your talk.*   As you prepare the title slide, start to think how you will use its features to start the talk and **introduce the subject**. Again, do not put too much effort into the regular weekly progress report, but for a major conference presentation, you can use the title slide to make some complimentary remarks about the occasion and the venue, and to thank the organizers for organizing and the audience for coming.

If you are giving a public lecture, you may wish to consider having the title slide displayed while the audience take their seats. You might even

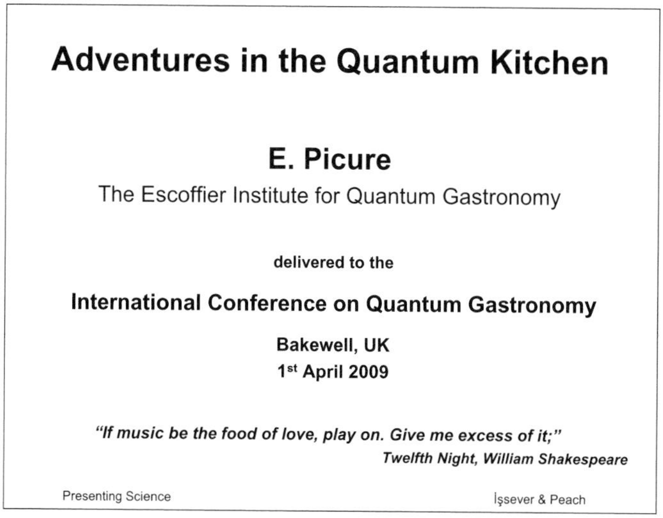

**Fig. 3.1** A simple title slide with the main elements—the title of the talk, the name and affiliation of the speaker, the occasion, location and date of the talk, and, as an option, a quotation giving some idea of the style of the presentation that whets the appetite of the audience.

want to entertain them a little, with a piece of animation, or (as with a colleague) an appropriate burst of the Grateful Dead. If this is really a public lecture, then try to remember the **copyright laws**! The point about such **unconventional beginnings** is that it has to be relevant to the lecture, and that the relevance is either obvious to the audience or explained on the title (or pre-title) slide. If you do wish to do something like this, turn up at the venue well beforehand and check that it works—nothing is less impressive than watching the guest speaker, the organizer, the hall manager and two graduate students fiddling about with wires for 10 minutes trying to make it work! The audience are thinking 'this person promises to tell me a fundamental secret of the universe, and yet cannot switch on a PC'. If you already have your Nobel, you might get away with it, but these days probably not.

*Introduction to the talk.* Whatever title slide structure you choose, eventually it will be your turn to speak. You may or may not have a formal introduction, ranging from 'I would like to invite Professor Wellknown, who needs no introduction, to deliver the address', to an embarrassingly fullsome recital of your CV since graduating. You need to connect the introduction to the talk, and this often needs a bit of improvisation. If the introduction is very brief, after thanking the organizers for inviting you to give the talk, you may need to say a couple of sentences about yourself, and why you find the subject of the talk so fascinating. If you are not fascinated by it, then neither is the audience likely to be. If, on the other hand, you have had the full CV treatment, you need say very little before moving into the talk, but you should nevertheless thank the session chairman for those kind words, perhaps commenting that when read out like that it always sounds like someone else, but, yes, you have indeed done most of those things, and then glide into your latest passion.

### 3.1.2 The outline of the presentation

After the title slide it is good practice to present the outline of your talk (see Fig. 3.2). Your talk is an entity which evolves in time and the audience needs to know what you will be presenting, and the order in which you will be presenting it, otherwise your talk will be most likely a roller coaster ride for them.

The level of detail in the outline of your talk depends on the length of your talk. In general if the outline contains more than two levels of bullet it will miss its purpose. If the talk is very long, one option is to present a

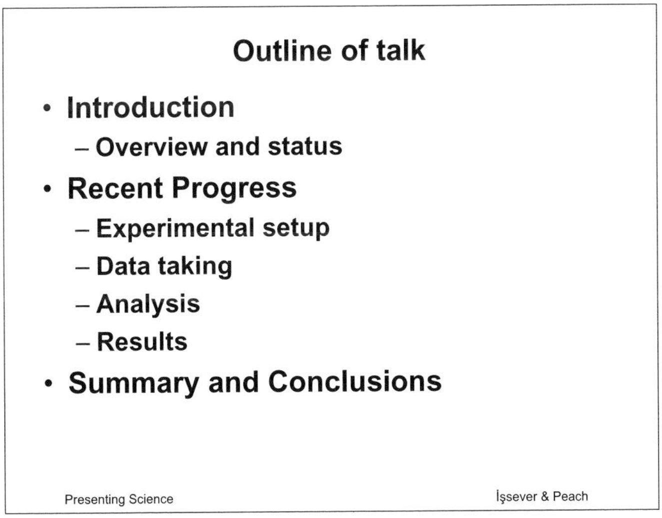

**Outline of talk**

- **Introduction**
  - Overview and status
- **Recent Progress**
  - Experimental setup
  - Data taking
  - Analysis
  - Results
- **Summary and Conclusions**

Presenting Science                    İşsever & Peach

**Fig. 3.2** A simple contents slide with the main elements and a single level of detail.

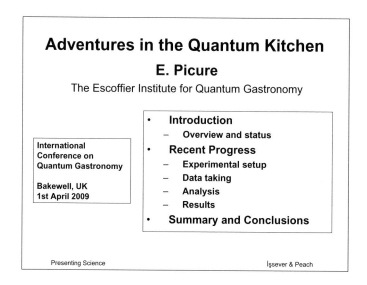

**Fig. 3.3** A combined title and contents slide.

more detailed outline when you reach a new section of your presentation, showing the listener in more detail what will be discussed next.

For short presentations, the outline of the talk could be included on the title page (see Fig. 3.3), saving time. Another time-saving tip is to use the outline slide as the basis for the introduction, talking briefly through each item of the outline (perhaps animating or emphasizing the text as you proceed), which works well if the subject needs little introduction and your main purpose is to explain why you have chosen to present the material in the way that you have. However, in general you should try to deal with the outline fairly briskly and move on to the meat of the presentation.

### 3.1.3 The introduction and motivation

The purpose of this part of the talk is to make sure that the audience knows **what** the talk is about, **why** it should interest them, and give them as much of the **background information** as possible so that they can understand and appreciate the significance of the work.

*Technical terms.* You need to make sure that any **technical terms** or **jargon** are fully explained—while you may have been working with the Advanced Diagnostics Dual Harmonic Quantum Wobbulator DHQW 450 for the last 10 years, at least some of the audience may not be so familiar with it. Always remember the graduate students! As a general rule, try to avoid jargon and TLAs,[2] although you may get away with IBM who, after all, more-or-less invented the TLA. The point is that jargon and TLAs are a private language, familiar to members of the club but impenetrable to those on the outside. Their excessive and unexplained use creates a barrier between you and your audience that will undermine your message. Of course, some technical terms are unavoidable, and you can gradually use them as the talk progresses with more confidence, but even if you use them on the slide, it is probably better to spell them out loud, at least until the audience recognize the rhythm of the language, each time you refer to them.

*Motivation.*  It is also a good idea to explain to the audience *why* they should bother to listen to you ... the **motivation for the work**. If you are unable to explain cogently and coherently why the subject interests you, you may find it quite difficult to convince them that they should take an interest. This need not take up much time, can be done in just one or two slides, but it should usually be there.

### 3.1.4   The technical or historical background

While you are of course familiar with the technical details of your work, and of the controversies or lacunae that you set out to resolve or repair, your audience may not be. You therefore owe it to them to give them the necessary background, succinctly and preferably in a reasonably objective way, so that they are in a position to appreciate the enormous contribution that you have so recently made to the subject. You can also use this to prepare the audience for your 'coup de theatre' by, for example, indicating where previous work had made assumptions that had subsequently been found to be unjustified.

### 3.1.5   The methodology

This section is very likely to be needed for a technical seminar, but should be brief (if present at all) for a general interest seminar. You can probably skip quickly over the standard items of equipment and methodology (no need to derive the least squares fit formulae), but should spend enough time on any new or innovative features of your work, so that the experts can understand what you have done. This is perhaps one of the

---

[2]Three-letter acronyms.

hardest sections for an inexperienced speaker to get right. For the young researcher, it is all new, and a great deal of time and effort has gone into understanding these points (in writing up the thesis work for example, and preparing for the viva). Neither the professional nor the lay audiences will, for different reasons, appreciate a long description of how a perfectly standard piece of equipment (say, a photomultiplier) works, unless, of course, your subject is how to make a much better photomultiplier for the laboratory, office and home.

### 3.1.6   The results

This is the heart of your presentation. This is what the audience has come to hear. You must not let them down.

*What is your key result.*   Before you can even start to prepare the slides, you need to know what your **key result** actually is. This may sound like a trivial remark, but it is not. How you present your results will depend upon many things. How surprising or new are the results? Are they controversial or just another stamp in the album? Have you been particularly clever or cunning in the analysis? What does it all mean? Do you (or we) know the significance of the result yet? These questions are of interest to the professional and the lay audience, but not in equal proportion.

*The audience.*   The **professional scientist** is a sceptic; we believe nothing until faced with no reasonable alternative explanation. Occam's Razor is one of our most powerful instruments. The professional scientist thus needs to be convinced that your result is sound. If it is controversial, the professional scientist will want a full discussion of the controversy, and to know whether there are any loopholes or dubious assumptions involved. The **lay audience** will want to know some of this, but will be much more excited by the significance of it all. They are likely to believe that you have done a good job (but be prepared to face some detailed questions afterwards), but really want to know what it means.

*The path to the results.*   Once you have decided what your result is, and how you wish to present it, then you have to devise a way of enabling the audience to share your sense of excitement and discovery. Not all of the steps in the argument are of equal weight or importance; these can be dealt with fairly briskly, especially if they are routine. However, the key parts of your analysis or theoretical speculations or whatever need to be explained in sufficient detail so that the audience (professional or lay) can follow. They will probably not be examined on the topic, and may not be asked to repeat it, but they need to see that you have taken all the right precautions, worried about the details, thought things through thoroughly so that there is no escape from your conclusions. If there is a lot of detail that just has to be presented, try to break it up into manageable chunks, with mini-summaries at the end of each chunk, so that when you come to put it all together for the big finale, you have taken the audience with you every step of the way.

*Presentation of the result.* Then, you arrive at your destination—the result itself. Think carefully how you are going to present this. In general, put in only as much detail as is needed to understand the result, and no more; neither you nor they want any distractions.

But you may be not quite finished. You may have to comment on the result, set it in the **context of previous results**—yours and other people's. You may wish to speculate, to a modest extent, about the next steps.

### 3.1.7 The summary or outlook slide

At the end of your presentation you should have a slide which reminds the listener what you have covered and what you want them to take away

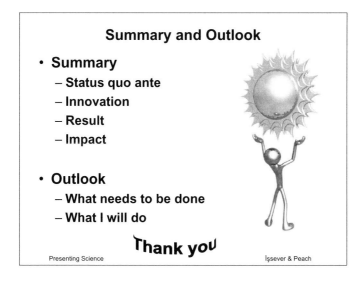

**Fig. 3.4** A generic Summary and Outlook slide.

from the talk. You repeat the important results, remind the audience briefly why they are important and also give the audience an outline of the next steps. The end of the presentation should not be dull. A typical final slide is shown in Fig. 3.4.

Finally, you may wish to thank the audience for their patience, acknowledge the contribution of your collaborators and co-workers (if not done on or after the title slide) and if appropriate invite questions. If you have done your job well, this should be the cue for thunderous applause.

## 3.2   The backup slides

A well-prepared speaker will not just a have good presentation but will also have thought ahead about possible questions the audience may ask and will have prepared some backup slides for this purpose. It always makes a good impression when you can show backup slides while answering questions from the audience. Sometimes if your time is limited and you are not able to go into much detail about a certain subject, you can put it into the backup slides and mention that you could give more details after the talk.

## 3.3   Summary

The structure of the talk is a key element in your presentation to enable you to deliver your message. We have discussed the different structural elements that usually make up a presentation: introduction/motivation of the subject, description of the technical background and methods, presentation of the results and finally the summary and backup information.

# 4
# Identifying the context of the presentation

*There is occasions and causes why and wherefore in all things*
*William Shakespeare (1564–1616), Henry V*

The context of a presentation determines, or should determine, how you approach its preparation. The context includes many things, the audience, the purpose of the presentation, the occasion, what precedes the presentation and what follows from it. It will define what you expect from the audience, and will influence how you prepare yourself for the talk.

*A simple example.* Suppose that you have been invited to give a series of lectures at a summer school. What more do you need to know, other than the topic? Here are a few of the questions that you need to have answered before you can start planning the course.

1. Is it an introductory course aimed at graduate students in their first year, or is it an advanced course more suited to graduates in their final year and young postdoctoral researchers?
2. Are the participants expected to ask questions during the lecture, or wait until the end?
3. Will there be any problem classes or discussion sessions?
4. Will lecture notes be handed out to participants before or after the lecture?
5. Will the proceedings be published, and if so, when?
6. What are the other lecture courses going to cover?
7. Will the basic theory already have been covered, or are they expected to know it already, or should you spend half of the first lecture going over it, just in case some have not seen it before?
8. If it is *your* job to give the basic introductory lectures, should you follow the standard approach in the usual text books, or should you

    assume that they have already covered that ground and try to give them more insight into the subject?

9. Will any of the lectures that come later in the school make any assumptions about what they have learned in your lectures?

10. Is there a social programme? If so, are you expected to participate in the activities and discuss the subject informally with the participants (which, from our experience, is always much appreciated), or can you spend most of the time in your room writing the next lecture?

Of course, in many cases the context is quite simple—you have been invited to give a colloquium on your work to colleagues in another university or laboratory, and they will be more than delighted if they have spent a happy hour listening to a good talk on a topic that interests them, and have picked up a few hints along the way that might help them with their research.

    The important thing to note is that the context is defined by the audience, the environment and the intentions of those who invited you.

## 4.1   Some common contexts

'Giving a talk' always has a context, which has implications for the way that you prepare your talk. Of course, much of this is straightforward and obvious, so if you are short of time, feel free to skip directly to the part that you are most interested in, or even skip it all. However, if you do get the context wrong, your talk may not be as well received as it deserves.

### 4.1.1   The group or staff meeting

In almost all scientific environments, there is some regular meeting to discuss progress, which we will refer to as the group or staff meeting. Here, your audience is composed of your friends, colleagues and potential rivals. It is an opportunity to develop your presentational style, perhaps with the help of a mentor. The key point about the context of this type of presentation is that you should be among friends. You can therefore skip most of the scene setting that is normally required. Nevertheless, your colleagues will appreciate a professional and businesslike approach. You should use the occasion to build or reinforce the group dynamics and to enhance your own reputation within the group. However, you should in general avoid too many in-jokes and 'amusing' asides, especially if these are made to disguise lack of substance or lack of preparation. Group meetings are work.

It is important to tell your colleagues what it is you are going to cover, and give them some idea of how long you intend to take. And in doing so, remember to speak clearly—there is a big difference between promising to speak for '4 to 5 minutes' and '45 minutes'. Except where the material is very new or very complicated, it is probably better to be brief, and expect them to ask questions if anything is unclear. **Simple, uncluttered slides work well**. If there are decisions to be taken, these need to be clearly identified, and the key facts upon which the decision rests need to be laid out. If the issue is likely to be controversial, it is probably a good idea either to have a written paper available as well, or to have more detailed **backup slides** (or both), to amplify the issues if the discussion drags on. It is also a good idea to let the others know beforehand that this is on the **agenda**, otherwise they will be unprepared for the discussion.[1] If you have a clear recommendation, make sure that it is clear; if there are doubts, indicate them. If there is a particularly tricky detail or a complicated equation, diagram or spreadsheet, again make sure that there are **paper copies** available—in fact, ask yourself whether you need a presentation at all. If you wish to encourage **dialogue** and **discussion**, consider giving the talk sitting down. This can be very effective with a remote mouse and laser pointer, since you are still part of the group rather than standing at the front lecturing.

Although you are among friends, you need to be alert. Revealing your great idea too soon could expose you to embarrassment if the idea is not that new at all, or run the risk, if the idea is really good, that others will take the credit. Waiting too long, and it may lose its impact—others with more experience may come to the same conclusions independently. It is here that you need to identify and have trust in your mentor.

**The key point about this context is that it should be informal, relaxed and efficient but not casual.**

## 4.1.2   The colloquium or seminar

The colloquium or seminar (in practice there is little difference—see section 2.1) is one of the main ways in which our research is communicated to, and discussed with, other experts in the field. It is also one of the ways by which we build and maintain the community of scholars. It is where we expose our ideas to informal scrutiny, and show off our flashes of genius. It is also the place where young researchers entering the field

[1]Never take a meeting by surprise, even with good news.

(postgraduates or postdoctoral fellows) are brought up to date about the latest developments.

You have been invited to give a presentation of your work because it is of interest to the department or group. There are many **motivations for the invitation**. You may have recently published an interesting result, and they would like more details. They may be about to publish an interesting result, and would like to find out what you know. It may be an area that they would like to enter, and need to know the state-of-the-art in order to assess whether it is worthwhile. You may be one of those they are considering for a new faculty position, and they want to check you out ('try before you buy') before proceeding further. Whatever the reason, they have invited you because they are interested in your work, and they will want to be 'informed, educated and entertained'.

Departmental colloquia can vary in size from a handful of people to an audience of a hundred or more, but in general they are relatively small (ten to thirty) and held in rooms that are also quite small. The introduction is likely to be a very brief extract from your CV, gleaned by the seminar organizer (a departmental duty for which there is often little competition) 5 minutes before the start, and then it is up to you. As noted in the introduction, you will need to know whether the audience are already experts in the subject, or whether you need to **set the context** and **review the field** as part of the graduate training programme. Getting this wrong can mean *either* most of the audience is bored for the first 20 minutes *or* half of the audience haven't the faintest idea of what you are talking about for the whole hour. Neither of these is a happy outcome. One way of addressing this, but which requires that you are confident of your ground, is to take a new approach to the standard introduction, or to make some perhaps controversial or stimulating comments. In this way, you can keep the attention of those who already know the subject well ('I had not really thought of it that way before') while bringing those new to the subject up to speed ('Hey, this looks like fun').

Your **style** should be informal—the audience will want some details, perhaps beyond what is normally available in the published paper. You should prepare **backup slides** with technical detail, in case you are pressed for it at the end. Your aim is to convince the audience (your peers) that you know what you are doing and why you are doing it.

There is an issue with 'work in progress'. You may not be ready to give the result (you may not yet have a result to give), but nevertheless you want to give your audience something. It is very important, if you are not able to give the result, that you make this clear in the introduction, otherwise you may raise expectations which are then unsatisfied, and the applause maybe more muted than your talk deserves. In compensation, you need to give more details, particularly of any tricks or clever ideas that have been uncovered along the way. Always remember, you are talking to a group of very smart people with between them a great deal of experience. If they do not know the facts, they can be of little help to you. There is of course a fine line to be drawn between giving them no information, which undermines somewhat the purpose of the talk, and giving so much information on how you achieve your remarkable successes that they steal the ideas and the glory.

There is an important variant of the colloquium, which is the '**General Interest Seminar**'. Most departments feel the need, a few times per year, to have a seminar on a topic and at a level of general interest, to which final-year undergraduates may be invited. Sadly, these are sometimes either not general or not interesting, and very occasionally neither. For the conscientious seminar speaker, these pose a particular challenge. In our view, it is much more important to be interesting than to be general. Undoubtedly someone in the audience will be expert in the subject (your name must have been suggested by someone) but most will not.

Here, you perhaps have three duties that you might wish to discharge. Firstly, you might want to **set the scene**—**why** is the subject interesting, **what** were the key discoveries and **who** discovered them. You should not be shy about your own contributions, but you should certainly refer to and discuss the contributions of others if it is appropriate. Secondly, you might wish to present the **current status**, and particularly your own contribution, in sufficient detail that the key breakthrough can be appreciated, but in less detail than would be appropriate for a research seminar. Thirdly, you might wish to speculate about the **medium- to long-term future** of the field. There is no 'golden ratio' for these three sections, but as a rough guide, spending **a third** of the time on the **first section**, **half** of the time on the **second section** and the remaining **sixth** of the time on the **third section** would give a reasonable balance. Be aware that, if you are generally interesting, you may receive more invitations to give such talks.

### 4.1.3 The conference or workshop

There are several different types of conference and workshop. A **workshop** is usually a small meeting on as specific topic, often with a '**what**

**should we do next**' feel to the agenda. A **conference** on the other hand tends to take a broader view, with the aim of assessing **where we are now**. Some workshops can be quite large (more than 100 participants); some conferences can be quite small (fewer than 100 participants). Participants in workshops are usually expected to contribute, whereas it is considered acceptable to attend a conference and just listen, although your institution may insist on you submitting a paper or a poster to qualify for financial support in attending the conference.

Perhaps more important, from the point of view of preparing your talk, than whether it is called a conference or a workshop is the size of the audience and the arrangement of the venue. For **very large audiences** (greater than 100), it is important that your *slides are clearly readable* form the back of the auditorium, and that you *speak clearly* with a steady voice. The formality of the occasion means that you are unlikely to be interrupted (unless of course you court controversy), but this also means that you receive rather little in the way of feedback from the audience.

For a **small workshop** (say ten to twenty people), the style of presentation is likely to be informal, with perhaps quite *frequent interruptions* for clarification or comment. These can be very stimulating occasions, and can lead to exciting new opportunities for further work (quite often this is an explicit aim of the workshop). Indeed, you may wish to encourage the active engagement of the audience during the talk, rather than waiting until the end for the questions. One way of achieving this is to deliver the talk seated rather than standing, so that the presentation becomes more of a conversation, but this is impractical for more than a dozen people.

We will refer to both as conferences in the discussion below, except where there might be a distinct difference in emphasis.

If you receive financial support from your group or department, you will often be expected to present a **summary of the conference** to the group or department on your return. Even if this is not a requirement, you should consider offering to do this anyway, to develop your skills in summarizing the work of others, and to build up an archive of useful material for when you are asked to give general interest seminars. When you become really well established, you may be asked to give conference summary talks, which are discussed below.

Conferences may be local, national, regional or international. As you develop your career, you will gradually receive invitations to speak at larger and larger conferences. Giving a good talk at a local conference will perhaps lead to an invitation to present a paper at the national conference, and so on.

There are different types of conference presentation, including **plenary session** talks, **parallel session** talks, **panel discussions** and **poster sessions**.

*Plenary session talks.*   If you have been invited to give a plenary talk at a major conference, the audience will expect you to be an expert. They will expect to be given new information or fresh insight, and will hope for both. You need to know whether you have been invited to **review** the subject (particularly taking into account papers submitted to the conference or presented in the parallel sessions), or to **present your own work**, or to give an inspiring talk **surveying the whole field**. This is your chance to make a big impression. *The challenge of the plenary talk is to challenge the audience with your insight into the subject while remaining within their 'comfort zone'.* If you pull it off, you can expect lots of invitations to give talks elsewhere. You should plan on spending a **huge amount of time preparing** and **rehearsing** this talk. This is also the opportunity to define your style—can you be both serious and amusing, can you make sweeping generalizations with confidence while accurately identifying the key concepts and controversies, are you in command, do you know all that is going on, is your judgement sound?

*Parallel session talks.*   The parallel session talk is usually about your work, occasionally a mini-review or meta-analysis. The **audience** has chosen to be there, and is likely to be interested and expert. If you happen to be involved in one of the hot topics of the conference, you might find yourself addressing an audience of sixty in a room designed for forty. On the other hand, the smallest room may hold 100, and only a dozen turn up to listen to you. You may not know how big the audience will be until just before your talk. Do not be too worried if half the audience leaves at the end of the talk before yours—they decided beforehand that there was another talk that they wanted to listen to (or they only wanted to listen to the previous talk). You should focus on those who stayed to listen to you—they had the choice, and chose to stay. Parallel session talks are often rather short, and so your skill as a speaker is challenged. You need to be **comprehensive** (otherwise you will be asked questions about the things that were not clear) but **brief** (otherwise you will run out of time before you get to the key result). This is one occasion where you might want to start at the end and work back towards the beginning—it is vital that you reach your destination, and the main variable is where you choose to start. You might also wish to give the 'Conclusions' slide immediately after the 'Contents' slide. While this does somewhat ruin the element of surprise, it does mean that you do present your result, and the audience will usually be happy that they know where they are being taken. Of course, you should be prepared for some of the audience to leave at this point, but do not worry too much—some of them will have just woken up and realized that they wanted to be somewhere else anyway, but will appreciate the fact that they have managed to bag one more result.

*Panel discussions.*   These are almost always disappointing, for both participants and audience. Each panel member is likely to be asked to make a short statement—do not go on too long, try to get your three main points in briefly and sharply, and hand on the microphone. The audience will then be able to comment and ask questions—again, try to be focussed and to the point. **Long, rambling answers will not impress**. Finally, and with little time left, you will be asked to summarize your point of view. Do not repeat your opening statement! The audience will be very impressed if you can address briefly some of the comments from the audience or the other panel members, but if not, wrap up quickly.

*Poster sessions.*   Poster sessions are becoming increasingly popular. Most participants will take a quick look and move on, but a few will stop and read carefully your poster—these people are your 'audience'. In general, you will not know them, but assume that they are influential—they might well be. Use all of your general communication skills, but with an audience of one! (See also section 5.7.)

*Conference summary.*   This is one of the greatest challenges, but (good news) it is one that you are only likely to face when you have made it as a great communicator. Done well, this can be very successful; done badly, it is embarrassingly awful—the entire audience is thinking 'I wish that I had caught the earlier train', and the conference organizers are asking 'whose idea was that?' **There are two basic approaches**. You can ignore what has been presented at the conference and give a stimulating and uplifting survey of the field. If you have something worthwhile to say, something fresh or daringly unconventional, you may get away with it. The audience are likely to wonder whether you and they have attended the same conference, but provided that they learn something new, and are entertained, the applause is likely to be warm and prolonged. On the other hand, if this is just your standard general interest seminar (see above), then they are likely to feel cheated, and the applause may be perfunctory. The other approach is to try to **summarize the conference**. This is hard work, and may mean that you miss some sleep the night before your talk, but it is very rewarding if you can pull it off, and the conference organizers will be delighted. If you accept the challenge, be prepared to be stressed, but also be prepared to be elated when you receive the 'I really enjoyed that' messages. This is one occasion when you will be forgiven for having far too much material. It is important that you **capture the key results** and that you **cover all of the plenary sessions**. The plenary speakers will usually be prepared to let you have advance copies of their talks, or an early draft, and will often provide guidance about which topics are particularly important. The audience will also appreciate your **personal perspective**, especially if clearly signalled, for example 'I certainly enjoyed the talk on Quantum Gastronomy by Professor E. Picure; while it is a relatively new development, and one that has yet to be absorbed into the mainstream, it provided significant food for thought'.

### 4.1.4   The grant or project proposal

These days, much of an academic's time is taken with filling in grant applications for funding, and occasionally these will need a presentation. Here, your **aim** is not so much to 'inform, educate and entertain' but to **convince**. The panel will not appreciate being bored or bamboozled but they will appreciate an efficient and comprehensible presentation of the proposal. This is serious business, and so fancy **animation** should be avoided. The slides should be **clean and clear**, and (most importantly) should address all of the issues that the panel have asked to be addressed. In the absence of guidance, the key issues are: **who is making the proposal, what the project aims to do, why you think it important, how you intend to achieve the objectives, and what is the timescale and what are the resources required to achieve the desired result**. Timing and fluency are important—use your team and other colleagues to **rehearse the presentation**, and listen carefully to their feedback—if you cannot convince them, you are unlikely to convince the panel.

When making a grant or project proposal, there is a very definite 'money slide', and that is the slide covering your **request for resources**. You will probably have submitted a written proposal, and you can assume that *most* of the panel have read it, and one or two will have read it very carefully. Even if the paperwork is clear, misunderstandings can arise; 'Do the salary figures include indirect costs? If they do, they look to be too low'. It is always tempting to 'cut and paste' the key financial **tables** (see section 5.4) from the written report, which has the advantage of being immediately recognizable to the panel and obviously consistent with the paperwork, but often has far to much detail to be absorbed quickly. On the other hand, a separate table, showing only key details may confuse. There is no easy answer, except to be guided by what will help your audience (the panel) to appreciate the strength of your case.

### 4.1.5 Presenting a report

The primary aim in the presentation of a report is to inform the audience about the subject—it may be a report on the progress of a project, or you may have chaired an enquiry into some issue of public interest or concern. There are several different audiences (for example, those who commissioned the report or a wider public presentation) that you may need to address. All will expect a comprehensive and straightforward account of the background to the report, the processes employed to gather the information and a concise summary of the principal conclusions. As with the grant or project proposal, the slides should be clean and clear, with few fancy flourishes. Recommendations and conclusions can be presented clearly and briefly, although any that are likely to be controversial should be explained.

### 4.1.6 The summer school or other academic lecture

Here the emphasis is on **education**. Some of the considerations were discussed in the introduction to this chapter. There are several others; for example, is the **material** easily available in a standard textbook—if so, you need to decide what extra the lecture brings to the student. It is said that the great theoretical physicist Paul Dirac used to read from his book when lecturing on quantum mechanics and, when challenged, would reply that he had put a lot of thought into the book and could think of nothing more that he could add! However, at least in our view, a lecture is somewhat different from a book—in a **book** the progression is **naturally linear**, whereas in the **lecture** it is possible to look both **forwards and backwards** to emphasize the importance of the subject, to identify features that will become more important, or which were identified as such earlier in the lecture. It is also possible to talk through the subject, perhaps bringing out unusual examples or applications, or referring to other areas where this particular topic has an impact. All of these things should help the student understand the relevance of the material, and to help in absorbing the content of the lecture.

On the other hand, if the subject is so new that it has not yet made it into the textbooks, you need to be much more **pedagogical**, or else you risk losing the students. References to the original research papers are, of course, necessary, but this may not be as helpful as you think, particularly to the less experienced students, since the research papers will be very likely to assume more general knowledge of the subject than they are likely to possess.

In planning the academic lecture, it is vital that you start with a clear idea of the key *facts* or *concepts* that you expect the student to have understood by the end of the lecture. These should be set out clearly in the course syllabus, and can vary considerably in granularity, from 'the derivation of $s = ut + \frac{1}{2}at^2$' to 'Newtonian mechanics'. However, in a standard 50-minute to 1-hour lecture, it is difficult to discuss in reasonable depth more than three or four topics, and if these are particularly advanced, rather fewer.

It is also essential to have a clear idea of what the *students* are supposed to be doing during your lecture—listening attentively, writing copious notes, jotting down the odd aside; this will govern the pace of your delivery, which also should influence the medium of delivery: blackboard, PowerPoint slides or writing on the projector. You also need to think carefully about whether to encourage questions during the lecture or not. With a small group, this can work very well, especially if you can provoke questions by asking **questions**. However, with a large number of students, this can be disruptive, especially if there are one or two who are very vocal.

### 4.1.7 The public lecture

The public lecture can be a great challenge and can bring great rewards. The public have chosen to come to hear you, and may even have paid for the privilege, and it is your duty to give them a good show. Most of them will not be expert in your topic, but all will be interested in learning more. They are quite likely to subscribe to a popular science magazine and to watch science programmes on the television. They may have done a science degree long ago, and still like to consider themselves as scientists. There may be members of the **audience** with a particular point of view, with a little knowledge and a lot of prejudice, especially if your subject is controversial, for example black holes or evolutionary biology.

In **planning your lecture**, you need to make sure that there is something for everyone. The **introduction** should be gentle, but defining clearly the subject so that the audience know where you are going. You need to set the topic in its proper historical and social **context**, and to separate clearly those things that are well established from those that are speculative. This is where your **oratorical skills** will be put to the greatest test. You need to **break up the talk** into sections (say about 10 minutes each), and give the audience time to reflect upon the information. They need **anecdotes** and asides, and **amusing details**, such as the time when the mouse came into the laboratory and almost fell into the liquid nitrogen.

It is good if you can use **props** (this is one of the detectors that we used on our first experiment) or **photographs** (this is me with the team opening the champagne when we first made the experiment work), to let the audience share your experience. Having something to give away (a leaflet or a poster) is always useful and appreciated.

At the end, you should expect **questions** (see section 6.9). These will vary enormously, from the very naive to the very sophisticated, and some definitely odd. If you lecture in public on any of the great scientific controversies with a religious dimension, you can expect at least one question about God; you should practise answering it.

Some people advise against starting your answer with 'that is a very good question', but we take the view that, if it is a very good question, you should say so. On the other hand, if it is a stupid question, it is probably better to avoid saying so.[2] If the question is quite weak, try to answer briefly, or try to formulate it in such a way that it becomes a useful starting point for a discussion. It is a good idea to **repeat the question**, even if there is a roving microphone, for the benefit of the audience. This has the benefit of giving you time to prepare the answer. Try to avoid getting into an argument with a member of the audience; in general the audience will be on your side.

Although not really a public presentation, dealing with the media is becoming increasingly important as a way of communicating to the public. We offer some advice on dealing with the media in Appendix C.

---

[2]Remember, there may be no such thing as a stupid question, but there is such a thing as a stupid answer.

### 4.1.8  The schools lecture

The schools lecture, where you have been invited to give a talk on a topic related to work in the syllabus, presents another special challenge. It has many of the same aspects as the public lecture, but you can get through even less material. It is important to be **enthusiastic**, and to give strong support to the teaching staff. With school students, you need to engage them through, for example, **eye contact**. Some will be there out of interest, but undoubtedly for some of them it will be seen as a way of avoiding lessons. However, some of the questions are quite likely to be interesting. Your aim is to keep those who are less interested entertained (so that they remain quiet) while trying to excite the small number who have a real passion for the subject, and who want to be informed and educated.

### 4.1.9  The job interview

The job interview can be a daunting event. For an academic position at lecturer level and above, it is quite often expected that you give two talks—one (half an hour to an hour) to the general department on your research and another (10–15 minutes) to the appointment board. For both of these, you should **keep strictly to time**—1 or 2 minutes less is far better than a minute more, but more than 5 minutes short should also be avoided. There is also a trend in some fields to conduct initial interviews by telephone or video conference; this presents particular challenges, for which you should carefully prepare (see section 2.4).

If you are asked to give a seminar on your research, then it should be treated as a normal academic seminar, but check carefully whether this is meant to be for an **audience** of experts or for a general interest seminar. Your performance will be judged, on both style and content, by both the appointment board and the departmental audience, who will very likely be asked to give informal feedback on the candidates' performance. They will ask themselves several questions. Is this research any good? What did he or she contribute to it? Is this a good seminar? Will she or he make a good lecturer (or is my workload likely to increase)?

The letter of invitation for the shorter presentation is quite likely to ask you to cover three things—**your career so far, why you want the job, and what you want to do if you are offered the job.** You need to cover all of the topics requested, but the appointment board is likely to be most interested in the third—what you will bring to the job; after all, they have already read your CV, probably several times. In general, go quickly through your early career, and stick to genuine highlights; it is much more important to present properly *your* contribution to your current work.

Appointment boards for academic positions (lectureships and chairs) can be quite large—ten is not all that rare—but for more junior post-doctoral positions the board is more likely to be three or four. However, in either case, you should be prepared for an **intimate style**, and try to engage all of the members of the board. This is an occasion where sitting down to give the presentation can work well, allowing you to use a conversational style.

Check whether a computer presentation is expected; if it is, remember that a small number of clear, simple slides is better than a large number of fancy, animated slides. Time is of the essence, and time waiting for pictures to float in is time wasted. If it is an oral presentation without slides, take even more time to prepare it; it might be a good idea to prepare for yourself a short pseudo-presentation with all the bullet points that you wish to get across.

It is often quite a good idea to have a handout to give to the board. If you have been asked to make a computer presentation, make sure that you have enough copies of the presentation to hand out to the board. If you have an oral presentation, try to get the key points on a single sheet of paper—the board will probably not have time to make detailed notes while listening to you, and they will appreciate having your summary to remind them of your strengths.

Unless you are very confident (and even if you are) it may be a good idea to go through your presentation with a good and trusted friend, and ask for their opinion. You have only one chance to get it right, so do not take any chances.

## 4.2   Assessing the audience

Once you have correctly identified the context, you can begin the task of assessing the audience (see also section 2.1). Now, we do not wish to overdramatize this step—it may take no longer than a few seconds. Most of the time, it is perfectly clear why the members of the audience have turned up to listen, and you can easily imagine their motivation. However, we do believe that you should do it, because otherwise you will not have considered the audience expectations, and thus you will have no real idea about whether you have tried to meet them or not. This reflects our view that the talk is there to meet the aspirations of the audience, not just to satisfy the ego of the speaker. Of course, the audience will have gone through the assessment of the context, probably unconsciously, just as you will have done.

The audience for a professional talk consists of fellow scientists, and attending your talk is part of their work. Departmental seminars are often held just before or just after lunch, so that members of the department (and the seminar organizer) can join you for a (free) lunch, or later in the afternoon, so that a few people can go on afterwards for a drink or dinner. Their expectation is that you will bring them up to date on the latest developments in your field, that is, work that has just been, or is about to be, published. Except for the general interest seminar, you can expect most of the audience to be reasonably expert in the field, but you should also take account of the needs of the graduate students, for whom this is also part of their training and who might not be so familiar with the broader aspects of the field, so that part of your talk may need to be rather pedagogical. In the general interest seminar, the audience are still knowledgeable, but for many it may be some time since they thought seriously about this particular topic, and there may be some final-year undergraduates or members of other departments who have come along because the topic looked interesting.

Assessing the audience for conferences and workshops depends critically on the type of conference or workshop. For relatively small meetings, it is quite likely that most of the audience will be real experts in the field, who will not appreciate a lot of time being wasted in going through the standard introduction to the subject. However, for large meetings covering a wide range of subjects, it may be necessary to cover some of the basic ground, albeit economically, especially if you are the main speaker on this particular topic. The experts will appreciate your insight into the subject, and the rest will appreciate being reminded why the subject is so important.

The audience for summer schools, and academic lectures in general, is very different, and what might be considered a good talk at a conference or seminar is quite likely to be unsuitable for a lecture. The students have come (or been sent) to learn, and you need to know something about their state of knowledge. There are many different learning styles—some people write copious contemporary notes, some listen intently and try to *understand* what lies behind the information, or some study the lecture notes later. It is very important that your lecture has a very clear and logical structure, with the important facts clearly identified and summarized. You also need to consider how to engage with the students, so that you can check that they are following your arguments, and that they have the opportunity to ask questions if something is unclear. You can use this feedback to develop a dialogue with the students that will help you pace the rate of delivery of the lecture optimally. If you have a series of lectures, you need to be prepared to revise later lectures in the light of the reception by the students of the earlier lectures, if you have underestimated or overestimated how much they actually know about the subject. If there is a wider range of knowledge (a mix of first and final-year graduate students, for example) you may need to consider holding an extra session to go through the basics, or to work through one or two examples.

The audience for a public lecture is perhaps the hardest to assess. Usually, they are there by choice—the title of your talk intrigued them, your subject is one that interests them, there is always a lecture in the local library on the first Wednesday of every month and they come along if they can, or colleagues from the local University have come along to see how you handle the subject. Try to keep to time—some may have a bus to catch, and rural transport is infrequent in the late evening. It is better to satisfy their curiosity over a limited domain than to mystify them over the whole territory. Try to anticipate some of their questions, and try to give them an answer to the question they are likely to be asked when they get home 'well, did you learn anything interesting?'

## 4.3   Judging the length of the presentation

Judging the length of the presentation is perhaps one of the hardest things to do. There is always the temptation to include too much material 'just in case'.

The first, and most obvious, thing is to find out how long the presentation is expected to be. Talks in parallel sessions can be quite short (10 minutes or so), or quite long (up to half an hour.) Plenary talks tend to be between half an hour and an hour, but check whether this is also expected to include **time for discussion** (i.e. is it a 30-minute talk plus 10 minutes of discussion, or a 30-minute talk including 5 minutes of discussion).

*How the size of the audience affects the length of the presentation.* In general, the larger the audience, the slower the pace at which you can present the information (except for the Conference Summary, see above). The rate of delivery can vary by a surprisingly large amount—perhaps

20%. A large audience is likely to have a broader distribution of expert knowledge, and so you will have to explain the key concepts more carefully. The psychological effect of the large audience also serves to slow down the presentation (we want to be heard at the back of the hall, and will tend to speak more slowly). However, a large audience is unlikely to interrupt you in full flow, whereas in a **small group** of half a dozen or so, you are quite likely to be asked questions during the seminar, which will introduce delay. You may be able to recover some of this by saying that the audience has used up some of the discussion time.

## 4.4   Timing

Timing is really only something that you can develop with time and experience. Unfortunately, rehearsal does not really help too much for the timing. The rehearsal may be in an empty room, or just to yourself, or to a small group of colleagues.

The best **advice** that we can offer, which we freely admit we often fail to follow ourselves, is to prepare the talk for the length that you have been given, and then *remove 5–10% of the material*. If you have overdone it, and have too much time, it is always possible to give slightly more detail on some of the later slides which, after all, should contain the key results, or remind the audience of where you are. Alternatively, and for other reasons, it might be useful to include some easy-to-digest slides (photographs of the apparatus, or of you showing a Nobel Laureate or the President around your laboratory) that are so obvious that they can be dismissed in a few seconds if time is pressing, or discussed in more detail if time allows. This also allows the audience to take a little rest, and prepare themselves for the next assault upon their critical faculties.

It is tempting to give advice such as 'assume that you can on average present one slide per minute', but this is misleading. As we noted above, some slides can be dismissed in a few seconds, and still get the point across. If the Pope dropped by to see your experiment, the audience will understand why you would like to show a picture of the event. Other slides, for example one with a complicated formula or the slide with your key results, may well take a few minutes to discuss properly.

There are some occasions where getting the timing *exactly* right is important. Making a presentation on **radio** or on **television**, especially if live, needs to be very precise, as are some rather **formal lectures** to some learned societies. As well as practising the talk, it is also essential to practise varying the **pace of the delivery**, so that you can speed up or slow down (by perhaps up to 10%) if you are behind or ahead of schedule. In a studio, there is always a clock, and the production team will be able to tell you whether to speed up or slow down. If there is no visible clock, try to arrange for a friend to give you signals every 5 or 10 minutes, or bring your own clock with you—but remember to switch off the alarm! For such occasions, it is useful to have mentally prepared two **summaries**—one very short, covering the basic message that you wish to get across to the audience, and a longer one which you can use if there is sufficient time, which can expand one or two of the more important points.

Finally, we need to consider your **state of nervousness**. In some people, this causes them to gallop through the presentation, arriving rather breathlessly to the conclusion several minutes early. In others, the need to make sure that everything is explained means that the first third of the presentation takes more than half of the time, and the remainder, which contains the main results and conclusions, is rushed. As you develop your presentational skills, you will become less nervous, or at least learn to live with nerves, and adjust the amount of material accordingly. But always remember—if you feel that you are rushing through the material too fast, the audience are likely to share your feelings and be less generous in their applause than your work deserves.

## 4.5 Am I the first speaker, or the last speaker, or somewhere in the middle?

If you are one of a number of speakers in a session, you need to be prepared. Keep strictly to time. The session chairperson should give you a clear signal when the end is approaching—for longer talks you are likely to receive warnings when there are 10 minutes, 5 minutes and 2 minutes to go, but shorter talks are more likely to have 5-, 3- and 1-minute warnings.

If you are the first speaker, you may be expected to provide a brief introduction to the session before continuing with your particular contribution; if you are further down the agenda, the introduction can be brief, leaving you more time for the key part of your presentation.

Being the last speaker in a session can be a challenge, especially if previous speakers have overrun their time. It is just not possible to deliver a talk satisfactorily in half the time, and so it is probably a good idea to have fewer slides and to be prepared to talk for longer if the time permits.

This will allow you to adjust to the time available without appearing hurried.

When you are part of a session or series of sessions have a look at the other presentations. Are there any talks that are likely to be relevant to your talk, or could your talk be relevant to other talks? If so, make a note to mention these connections, and refer to them briefly during your presentation.

## 4.6 Summary

Identifying the context of the presentation is a essential part of the preparation of the presentation. Many factors need to be considered: What is the type of event the presentation will be given at? What are the expectations of the audience? What is their level of knowledge? How much time do you have for the talk?

# 5
# Style

*One man's style must not be the rule of another's*
*Jane Austen (1775–1817), Emma*

The style of the talk should be chosen such that it supports the message, the occasion and your personality. The guiding rule should always be that the chosen style should not make the slides hard to read or distract the attention of the audience from the message. For example having an **animation** on the slide which runs all the time will most certainly catch the eye of the audience, because our eyes are programmed to look at moving objects. But this will make it impossible for the listener to follow your discussion on the rest of your slide while this animation is running. You can have different styles for different occasions. Figure 5.1 shows an example of a slide style for a working group meeting and in Fig. 5.2 for an invited seminar talk on the same topic.

Style features which are important are the **font style**, the **slide background** and **border**, **colours**, **plots**, **tables**, **pictures** and **animation**. We will cover each of these in the sections below.

## 5.1   Font

The choice of the font should be a conscious decision and should not be just guided by your aesthetic feelings or the system default. There are several choices to be made under the font banner, and all are important.

- Style: Serif or Sans Serif,
- Appearance: normal, **bold**, *italic*, underlined or ***combined***,
- Size: small (8 pt), medium (12 pt), large (16 pt), huge (20 pt), vast (24 pt),
- Colour: foreground (text) and background.

*Colours.*   Colour theory is quite complex, and beyond the scope of this book. However, a few simple ideas should help avoid most problems. The colour of the font and the colour of the background need to be considered

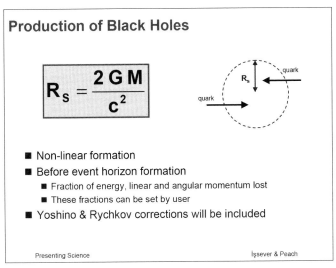

**Fig. 5.1** Example of a slide for a working group meeting. Please refer to plate section for colour figure.

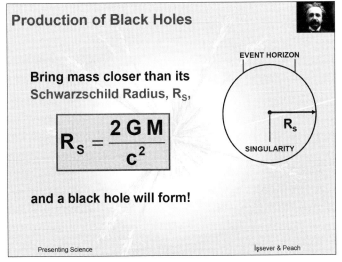

**Fig. 5.2** Example of a slide an invited particle physics seminar on the same topic as figure 5.1. Please refer to plate section for colour figure.

together; it is better to choose colours that are readily distinguishable, with high contrast. A yellow (black) text on a white (dark) background is nearly invisible (see Figs 5.3 and 5.4). Remember that your presentation may be printed on a black and white printer and colour-coded information may be lost if there is low contrast, and that between 5% and 8% of males

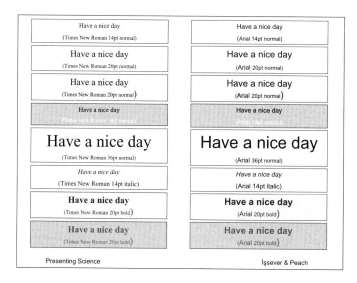

**Fig. 5.3** Different text font styles and background colours. This slide shows examples of serif (left column) and sans-serif (right column) fonts. Please refer to plate section for colour figure.

and less than 1% of females are **colour blind**.

*Font type.* In general we think that you should not use more than one font type in a presentation. If several font-types are used, it is just distracting, and one more thing that the brain of your listener has to filter out or deal with. The best font styles for presentations are in our opinion sans-serif fonts. Serif fonts appear cluttered on the screen and are tiring for the eyes.[1] Because of this you also should have a very good reason to use italic fonts. You can use bold or underlined appearance to highlight text. The best font style in our opinion is 'Arial', although 'Lucinda Sans' also looks quite pleasing. The size of the font will depend on whether it is the title or text on the slide, but the bigger you can make the font size the better it is. Try to use all the available place on the slide if possible.

Figures 5.3 and 5.4 give some examples of text fonts and background colours. What do you like?

*Fancy fonts.* If you use any fancy fonts (Fig. 5.4), you must embed them with the presentation, otherwise you risk having a very different layout

---

[1]Very different considerations apply for written text—this book is typeset in a Computer Modern, which is a serifed font which both we and the publishers think looks very pleasing to the eye on paper.

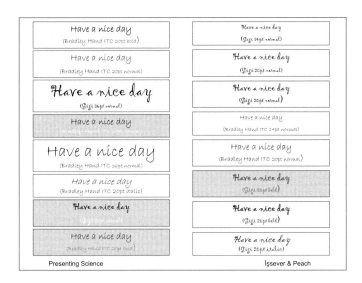

**Fig. 5.4** Different text fancy font styles on different background colours. Please refer to plate section for colour figure.

(Greek not turning out as Greek, for example, makes a real mess of formula or Feynman diagrams). For a PowerPoint presentation, we recommend that you use OPTIONS/SAVE/EMBED TRUETYPE FONTS.

There are some occasions where using a different font is helpful. For example, if you are quoting someone else's words, it might be a good idea to use italicized Times Roman in quotation marks, just to emphasize that these are not necessarily *your* views—the font change is a clear visual signal that these words are 'foreign' to the presentation.

## 5.2   The frame

We will refer to the combination of slide background, slide title and any other information that appears on each slide as the slide *frame*; in PowerPoint, this is referred to as the *slide layout* (VIEW/MASTER/SLIDE MASTER). Using a frame (like those in Figs 5.14–5.17) gives the talk a coherence. Try developing your own frame, which you can use again and again. This will allow you to re-use slides with the minimum of effort. In PowerPoint, this is most conveniently achieved by developing a standard *template* (FILE/SAVE AS Save as Type: Design Template).

The frame should usually contain a **slide title**, somewhere near the top of the slide and summarizing the slide contents—this helps the audience follow the development of the talk. It is also useful to have a footer containing, mostly in rather small type, your name, perhaps your institution, the date and the number of the slide in the presentation. This identifies the slide as yours and (through the date) the occasion. (When you next give this talk, remember to change the date!) This also encourages those who cut and paste from your talk to give you some of the credit.

The **slide number** should be readily visible and is useful for the audience to note if they have questions at the end of the presentation. (For example, 'I wonder if you could go to slide 22 and explain again why the peak at the extreme left can be ignored?') As we have noted elsewhere, if your presentation is via video-conference or over the telephone, *you must have slide numbers.*

Beyond this is really a matter of taste.

You may wish to give the audience an idea of where you are in the talk. One way of doing this is to add the total number of slides to the slide number ('slide 22 of 49') but this can be daunting if you have a large number of slides or backup slides. Some packages (but not PowerPoint) divide the presentation into sections, which are displayed across the top of each slide, with an indication of the current section and the slide within the section.

It is quite common to see, usually in the top left or top right corner, the institutional logo, and even a whole battery of logos (University, Department, Group, Experiment ... ). As with everything else, if it serves a good purpose, then do it ... but we would advise identifying the purpose. And similarly with other embellishments.

### 5.2.1 Background

The background of your slides can, if chosen wisely, help you to get your message across. That said, we often see examples of poor background choices because the speaker was more guided by presenting an artwork as the background choice than supporting the main function of a talk: getting the message across.

Poor choices for backgrounds are in general those with multi-colours or with a lot of features on them, see, for example, Figs 5.5–5.13. Multi-coloured backgrounds have the disadvantage that the font colour you have chosen will not work at every location on the slide. If you chose a photo to be the background of your slides, for example a picture of your detector, you should try to blend away the features of the photo by editing it with any modern graphic programme to adjust the brightness and contrast (see

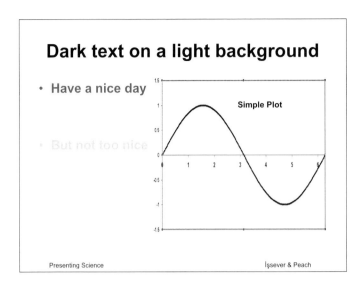

**Fig. 5.5** An example of a combination of a light-coloured background and different font colours. The green and the yellow font colours are not easily readable. Please refer to plate section for colour figure.

Figs 5.12 and 5.13). The level of transparency needs to be chosen carefully and you should always look at your slides with a projector on a screen and not just on your computer screen, because often there are significant differences between the colours displayed on the computer screen and those projected onto the screen of the lecture theatre. This arises from the different way in which the colours are generated and combined in the two technologies. There is a similar problem with the overhead projector, where it is essential to make sure that the printer is told that it is printing onto transparencies, otherwise the colours look correct in reflected light but not in transmitted light.

If you want to get away from a plain white background, which can be quite hard on the eye, you could try a pastel shade such as a very light blue or pale yellow, or perhaps two rather similar colours merging from top to bottom, for example a light blue at the top and a blue with a slightly greener tint at the bottom, which will be restful on the eye, but not limit your choice of font colours significantly.

## 5.2.2 Borders

Borders can add a lot to a slide, but if chosen wrongly they can also distract from the main body of the slide (see, for example, Fig. 5.15).

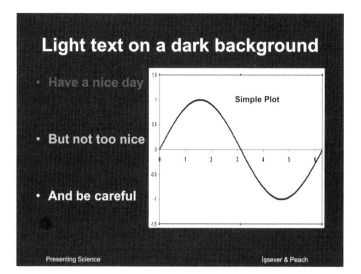

**Fig. 5.6** An example of a combination of a dark background colour and different font colours. The white, yellow and green fonts are clearly visible whereas the red font is very hard to read. Please refer to plate section for colour figure.

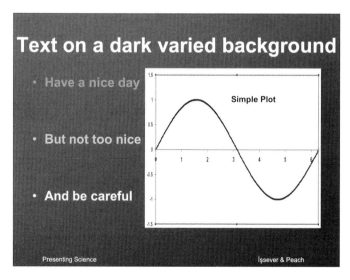

**Fig. 5.7** This figure shows a slide with varied background colours and different font colours. The white, yellow and green are readable but the red font is barely visible. Please refer to plate section for colour figure.

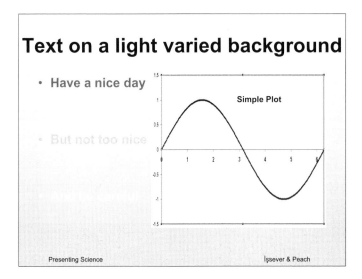

**Fig. 5.8** This is another example of a combination background and different coloured fonts. The yellow and green fonts are not a good choice whereas the dark brown is readable. Please refer to plate section for colour figure.

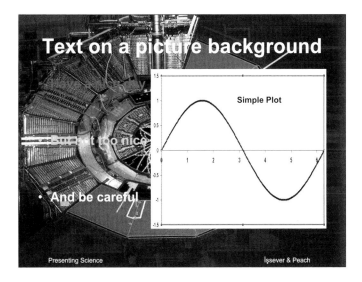

**Fig. 5.9** This is an example of a slide where a photograph was used as a background. The text on this slide is very hard to read. Please refer to plate section for colour figure.

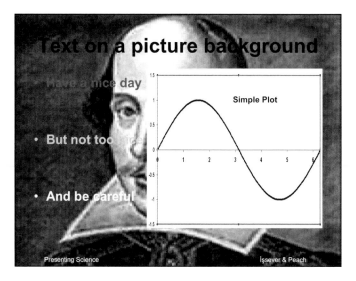

**Fig. 5.10** This is another example of a slide with a photograph as a background and again the features of the photograph are making it very hard to read the text on the slide. Please refer to plate section for colour figure.

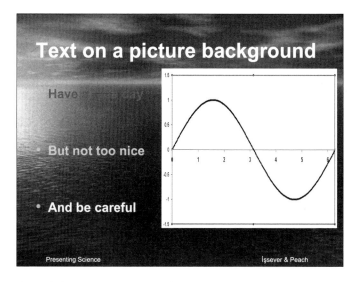

**Fig. 5.11** This slide uses a photo of a lake as a background which distracts from the text and makes it hard to read. Please refer to plate section for colour figure.

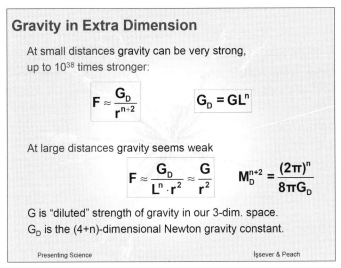

**Fig. 5.12** Slide showing a background picture which was modified such that it is not disturbing the readability of the slide. Please refer to plate section for colour figure.

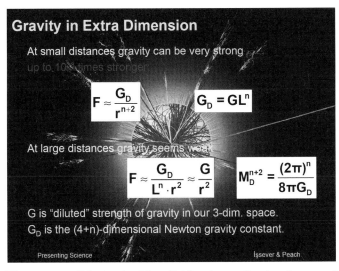

**Fig. 5.13** The same slide as in Fig. 5.12 where the background picture was not modified. Not all the font colours are working well with this background. This slide shows that one way to make text readable on a busy background is to use filled boxes around the text you are displaying. Please refer to plate section for colour figure.

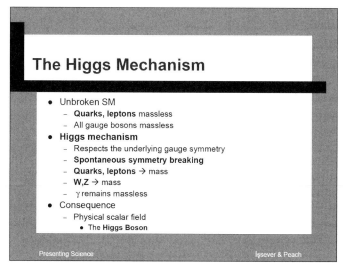

**Fig. 5.14** This is an example for a border style which uses a lot of space. Please refer to plate section for colour figure.

Another point to keep in mind when choosing a border style is the amount of space occupied. Space is precious and if you have the choice between making the font size larger or choosing a fancy border style taking 20% of the total frame area (forcing you to use a smaller font size), opt for the style which does not compromise your font size. Examples of different border styles are given in Figs 5.14–5.17. The first two examples (Figs 5.14 and 5.15) show examples of border styles which are wasting too much space, whereas the two later slides (Figs 5.16 and 5.17) are examples of a more economical style.

There was quite a fashion some years ago for using a bright yellow font on a dark blue background, and we certainly agree that the text really stands out. However, we are not convinced that this is really appropriate for a scientific presentation; it looks a bit like trying to convince by shouting rather than by persuading. Likewise, some people use a pseudo-chalky blackboard (neither black nor white) as a background, and use a pseudo-handwriting font like Comic Sans MS in white bold to produce a facsimile of the 'chalk-and-talk' style, but this can also get in the way of the message—we *know* that it has been all carefully prepared beforehand, and that it is not spontaneous. But, if you like it and it is consistent with your self-image, why not?

There is one place where a more flamboyant background might work well, and that is on the title slide. Here there is usually rather little text,

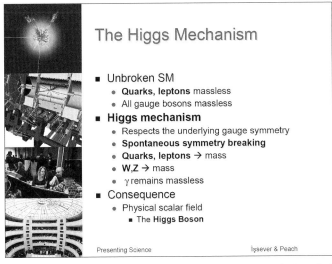

**Fig. 5.15** This border distracts from the main body of the slide and it also takes too much space. Please refer to plate section for colour figure.

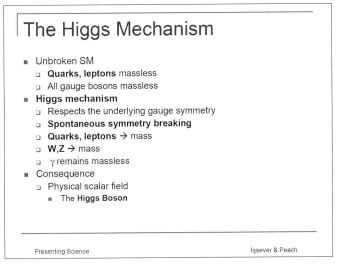

**Fig. 5.16** This is an example of a more economical border style where there is a lot of space for the main body of the slide left. Please refer to plate section for colour figure.

and so it is possible to arrange what text there is (see section 3.1.1) around the features of the background, and help prepare the audience for what will follow, especially if the title slide is displayed while the audience take their seats.

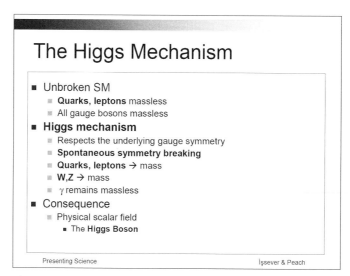

**Fig. 5.17** This shows another example of an economical border style. Please refer to plate section for colour figure.

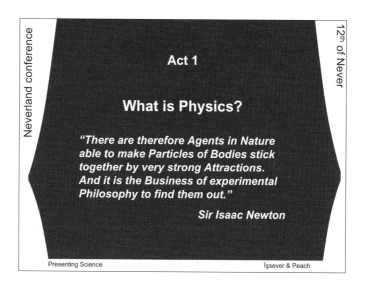

**Fig. 5.18** An final example of a combination of background and font colours and a 'theatre curtains' border. Please refer to plate section for colour figure.

Finally, you might want to combine the background and the border into a frame that sets the stage, as in Fig. 5.18

## 5.2.3  The corporate frame

A particular issue for scientific presentation is the 'Corporate Style'. The main purpose of the corporate style is to sell the corporation, and as a result the corporate image tends to loom large while the space left for the message is rather small. Figure 5.19 shows a typical, professionally

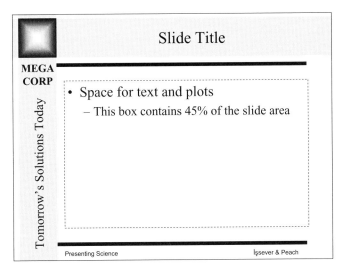

**Fig. 5.19** Example of a corporate frame.

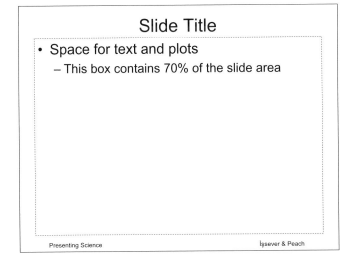

**Fig. 5.20** Example of a simple scientific frame.

produced, corporate style. Note that only around 45% of the slide area is available for the message, because the message is anticipated to consist of three or four short bullets. You know the sort of thing—slide title: 'Vision'; slide contents: three bullets, 'Be the Best', 'Beat the Rest', 'Beat the Breast'. Actually, this style, with its fussy regular serif font (Palatino) carries rather little weight. A simple scientific frame is shown in Fig. 5.20, where around 65% of the slide is available for content. The problem with the assertive corporate style is that it is also intrusive, and after a while annoys the audience. However, if your employer insists that you use it, there is not much that you can do. Otherwise, our advice would be to see if you can get away with using the corporate style for the title slide, the contents and the conclusions slide, and use something more appropriate for the bulk of the talk.

## 5.3 Bullets and boxes

When writing a paper or an article, usually most of the words go into the text—phrases, sentences and paragraphs. Sometimes, we may need to make a list of things or introduce some numbered points, so that we can refer to them later. Occasionally, we may need to make a remark which does not quite fit in with the flow of the text, but which is important for the reader to understand; if the thought can wait, it can be relegated to an Appendix, but if it has more immediacy it can be placed in a separate box—a text figure if you like. Similarly, when we give a talk, there may be occasions when there is a list of things that we want to bring to the attention of the audience, or make an aside that adds to the general development. But the use of bullets and boxes in a talk goes very much further.

Bullets and boxes are used to present the main points, or to add detail to a plot or a picture. Used judiciously, bullets and boxes can really help the audience understand the content.

There are no real rules about when to use a box or when to use a bullet, but the following may be helpful in deciding. **Boxes** are most useful in providing additional information, or commenting upon the topic; typical examples might be (see Fig. 5.21) identifying the components of a picture or plot, or attributing a source, or adding a restriction. **Bullets** on the other hand present a logical argument or development and, through indenting, can group thoughts together, as in Fig. 5.21. Although there are very few words on this slide, and only one simple plot, it contains a lot of information. Depending upon the occasion, you may need to spend a

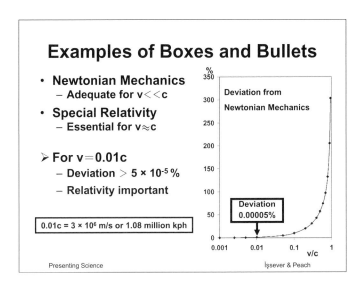

**Fig. 5.21** An example of the use of boxes and bullets. On the left is a series of bullets developing the limits of validity of Newtonian mechanics, and the criterion for changing to special relativity. The plot on the right shows the percentage deviation from Newtonian mechanics as a function of the velocity $v$ relative to the velocity of light $c$. The box above the arrow indicates the position where this deviation exceeds 0.00005%. The box on the lower left translates the limit of validity indicated by the box in the plot to a velocity, expressed as both m/s and km/hour.

rather short amount of time on this slide, if the audience can be presumed to be generally familiar with the subject. On the other hand, this could be the first slide of a section which amplifies and explains the difference between Newtonian and Relativistic mechanics. Your thoughts about this slide might be something like the following.

*Newtonian Mechanics—Newton's three laws: (1) A body remains at rest or moves in a straight line at constant speed unless acted upon by an external force (2) the acceleration (the rate of change of velocity) is proportional to the applied force, and acts along the direction of the applied force and (3) to every action there is an equal and opposite reaction. For most applications, where the velocity is very much less than the speed of light (300 thousand km/s), this is a perfectly adequate description of mechanics. However, when the velocity approaches that of the speed of light, Newtonian Mechanics is no longer adequate, and we need to use* Special Relativity *as formulated by Einstein. The Michelson–Morley experiment had shown that the velocity of light was independent of the direction of motion of the measuring apparatus, and Einstein realized that this implied that the velocity of light (which had by then been measured) was a universal upper limit, and that Newtonian Mechanics needed to be modified. (We will look at the modifications to Newtonian mechanics in the next slides.) The plot on the right shows the deviations from Newtonian mechanics as a function of the velocity relative to the velocity of light. When the velocity is about 1% of the velocity of light (that is, about 1 million kph) the deviation from Newtonian Mechanics is about 0.00005%. Whether this is significant or not depends upon the accuracy with which the measurements are being made. In general, Newtonian Mechanics will still be quite adequate, but (for example) for the most accurate measurements of time using atomic clocks, Special Relativity must be taken into account. In fact, for extremely accurate measurements, and for the correct functioning of satellite navigation equipment,* General Relativity *must also be taken into account.*

You might also wish to discuss the similarities and differences between the concepts of energy, momentum, mass and velocity in the relativistic and Newtonian regimes, and their dependence upon the frame of reference.

As you take your audience through the slide, you can highlight the part that you are currently discussing by either reveal animation or judicious use of the pointer.

How you use bullets and boxes is up to you. In the examples shown in Fig. 5.21, the bullets and the plot contain the important messages; the slide would be complete without the boxes, and the **boxes contain commentaries** on the important messages contained in the bullets and the plot; without these, the boxes would be difficult to interpret. Note also the change of bullet style for the third bullet, reflecting the fact that this is a *specific instance* and not a general point. However, it would be equally possible to emphasize the key messages by putting them into boxes, and to use bullets to deliver a commentary.

It is very important to understand the difference between writing a sentence or paragraph in a journal article and writing text for a bullet or box. Compare the information in Figs 5.22 and 5.23—which would be easier to absorb from the back of the auditorium? In bullets and boxes, it is good to have a '**one thought on one line**' policy. If the thought spreads over more than one line, then the members of the audience have to read to the end of the line and then scan back to the beginning and

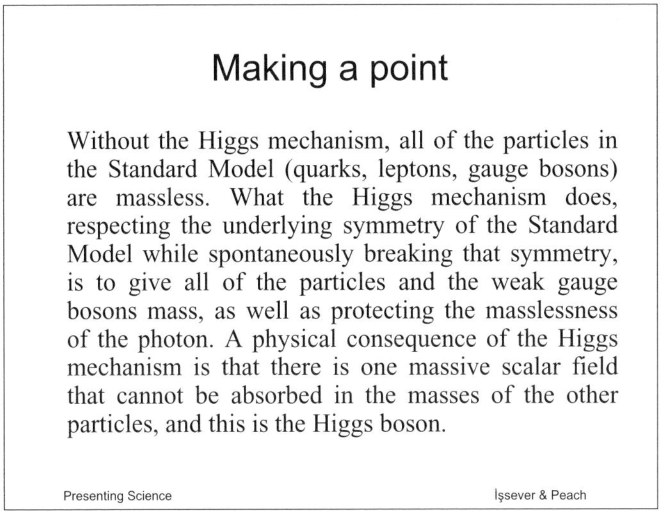

**Fig. 5.22** In this example, the slide has been written using complete sentences, as in an article or book. The key message is difficult to see, and from the back it may be difficult to read, especially if the slide is not displayed for very long.

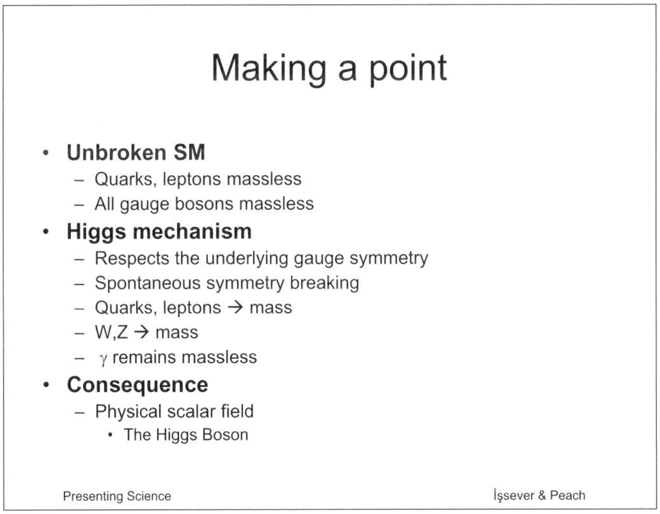

**Fig. 5.23** In this example, the same information as in Fig. 5.22 is given, but using bullets. It is now relatively easy to see the key facts and follow the argument.

carry on reading; this takes time and effort, especially if you are near the front. It can be annoying if the 'reward' for making the effort is rather small, for example, attaching the 'GeV' to the 'Energy=10'. There are several techniques for getting the thought on to one line.

- Try removing all unnecessary words, like 'the' and 'is', and unhelpful adjectives ('very' in front of 'large') and adverbs ('violently' after 'exploded').
- Try using shorter words ('red' rather than 'vermillion') or using standard symbols ('E' rather than 'Energy').
- Try splitting the thought into two half thoughts (instead of 'Calorimeter Resolution $\sigma(E)/E \simeq 0.01 \oplus 0.08/\sqrt{E} \oplus 0.005/E$' put the thought on two lines, with 'Calorimeter Resolution:' on one line, and the rest on another line, preferably indented.
- Try expanding the box a little sideways, if the final word is quite short (like the 'GeV' above).
- If all else fails, try a slightly smaller font size; it is a matter of judgement whether it is more difficult to see the smaller font or to deal with the unfortunate line break.

**Bullets** allow you to present the *essence* of the argument economically. In written or spoken language, there is a lot of padding to slow down the rate at which information is transferred so that we can absorb it. Consider the following sentence.

After calibrating the response and correcting for dead channels and inefficiencies, 235 events remained in the signal region, and 105 events in the side bands used to estimate the background, giving a signal of $130 \pm 19$ events (statistical error only).

How much of this sentence is needed to make sense of it? Not much! The key words are *calibrating, correcting, signal, background* and *statistical errors only*. The key facts are the four numbers given—235 total number of events, 105 background events, 130 signal events and an estimate of the error on the signal of 19 events. This information can be presented simply in a few bullets.

- 235 events
  —after calibration and correction
- 105 events background
  —in side bands

- $130 \pm 19$ signal
  —statistical errors only

Of course, this would also be made much clearer by including a plot. Note that although there are fewer words in the bullets than in the sentence (18 words and numbers to compared with 40), the bullets occupy more lines (6 to be compared with 4). However, the ease with which the information can be absorbed by an audience is much greater in the bulleted version.

When using bullets, it is always a good idea to use a **bullet symbol** for each line, with different bullet symbols and indentation for different levels of bullet. The bullet symbol is the visual clue that this is a new thought, just as the initial capital letter after a full stop signals a new sentence. The demarcation for the end of a bullet is the new line if, that is, you follow our advice above and force the coherent thought into a single line, just as the full stop marks the end of the sentence; if not, the thought continues until the next bullet symbol.

Visual clues about the level of importance of certain words, or the

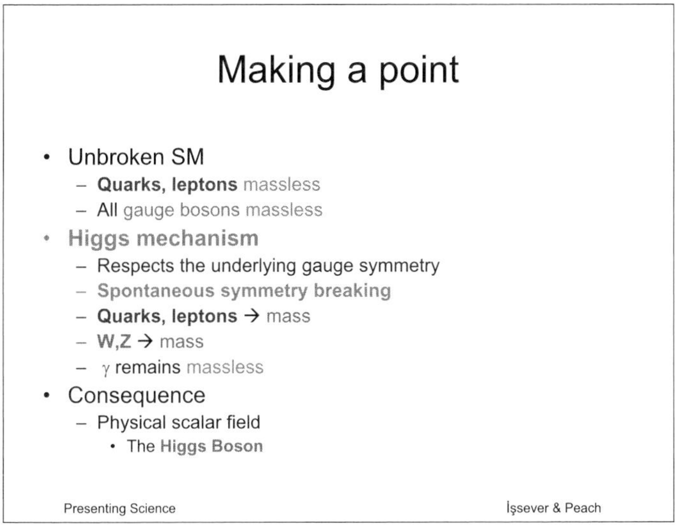

**Fig. 5.24** In this example, the same information as in Fig. 5.23 is given added emphasis using colour. Blue and purple are used to distinguish between *fermions* and *bosons*, red is used to highlight the problem that the *Higgs mechanism* addresses (*masslessness*) and the main feature of the mechanism (*spontaneous symmetry breaking*) and green links the acquisition of *mass* by the fermions and bosons to the *Higgs Boson*. Please refer to plate section for colour figure.

relationship between them, can be enhanced by the use of colour in bullets as in Fig. 5.24. In general, it should be possible to arrange thoughts so that no more than three levels of bullets are required, in which case three bold colours such as black, blue and red should be sufficient.

## 5.4 Pictures, plots and tables

*Pictures.* Pictures serve many purposes in a presentation. They can be used to give the listener a break or they can say more than the proverbial 1000 words about a complicated piece of apparatus or experimental site. Pictures can enliven an otherwise pedestrian presentation, but can become tedious if this reduces to the level of 'here are some snaps I took while on holiday'. A montage can look quite impressive, and help give the audience a real 'feel' for the issues. The eye/brain combination is a fantastically fast 'pattern recognition' machine, and so you do not need to spend a long time on straightforward pictures. However, for each picture shown, you should at least indicate with a caption, a title or in words, what is shown and why you have shown it.

*Plots.* Plots, on the other hand, are the very heart of a scientific presentation (and yes, that includes theory!) It is very difficult to think of any subject that would not be made more clear to the audience than a well-chosen plot. Given their importance, then, it is surprising how often plots fail to make the impact that they should. Often, this is because one or more of the following obscure the meaning:

- the plot is too small;
- there is no title;
- the axes are unlabelled, or the labels are unreadable;
- the scales are unreadable or missing, as are the units;
- there are a lot of extra 'lines' that are irrelevant to the point being made;
- the plot was originally in colour but is now in black and white;
- the plot itself is meaningless (many 'double log' plots are meaningless, as are many 'data–montecarlo' comparisons).

Figure 5.25 shows an example of a slide where the figures and the labels of the figures are too small. The figure on the left hand side of the slide has too many lines and will certainly confuse the audience. We would recommend showing each the figures on this slide on separate slides, or using animation to enlarge the figure while you are discussing it and to

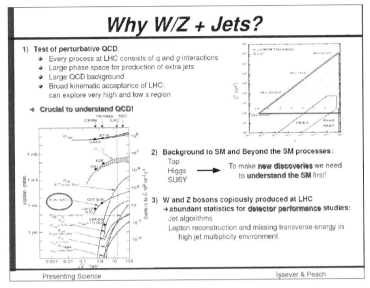

**Fig. 5.25** Example of how not to present figures. Please refer to plate section for colour figure.

minimize it once you have finished (see Fig. 5.30). The figure on the left hand side needs careful editing so that only the relevant information on it is shown.

Sometimes, it is not possible to change the appearance of the plot—it is taken from another source and cannot easily be re-drawn to make it more visible. There are several things that can be done. If the axes or labels are not clearly visible, you can try overwriting with thicker lines and new labels in a more readable font. It is more difficult to manipulate the fitted curves or data points, but here a properly used pointer can help the audience see what might be indistinct.

We also would like to make you aware of issues with graphics and file sizes. The size of the presentation file can become very large if you use many high-resolution pictures and figures, which can make the uploading a talk to an agenda page via a web browser very slow or sometimes even impossible. Colleagues to whom you send the presentation by email may not appreciate it when their inbox overflows or their email system breaks down because of large size of the attachment. If at all possible you should try to compress the figures as described in Appendix B.3 or convert your PowerPoint file into a pdf file (but be aware that this may be a problem if you are using animation).

*Tables.* Tables need to be carefully presented because there is always a temptation to put too much information onto the slide. Most tables presented in talks are just a large number of numbers and the listener does not know which of the many are important and which are there because the presenter could not be bothered to edit them out. If you want to present your results in a table, try to make sure that you show only the rows and columns which are relevant for the discussion. Make sure that the font size is big enough so that the numbers are readable, even from the back of the hall. If possible highlight a few key numbers with boxes, different font colours or font styles.

We do not recommend cutting and pasting tables from publications. If you have to do it, make sure that you use boxes or equivalent tools to indicate to the audience the important bits of information in this table. If you want to go through the whole table bit by bit, use animation to indicate where you are in your discussion with regard to the columns or rows in your table.

Figure 5.26 shows an example of a table which is not too dense and where a red box is used to highlight the row the speaker is currently discussing. Figure 5.27 shows a table which was taken out of a publication and the important columns are highlighted with differently coloured

### Sample Table

| Ms [TeV] | $M_D$ [TeV] | $M_{thresh}$ [TeV] | $\sigma$ [pb] |
|----------|-------------|--------------------|---------------|
| 1.0 | 1.5 | 3.0 | $(2.32 \pm 0.01) \times 10^{+1}$ |
| 1.2 | 1.8 | 3.6 | $(4.70 \pm 0.01) \times 10^{+0}$ |
| 1.4 | 2.1 | 4.2 | $(9.58 \pm 0.01) \times 10^{-1}$ |
| 1.6 | 2.4 | 4.8 | $(1.87 \pm 0.01) \times 10^{-1}$ |
| 1.8 | 2.7 | 5.4 | $(3.34 \pm 0.01) \times 10^{-2}$ |

Presenting Science     İşsever & Peach

**Fig. 5.26** Example of a table which is not too dense, where a red box is used to highlight the row under discussion. Please refer to plate section for colour figure.

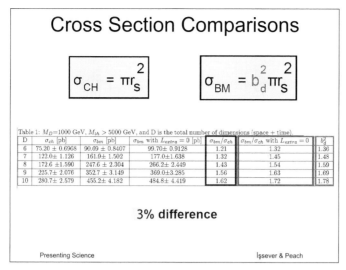

**Fig. 5.27** Example of a table taken from a publication, with important columns highlighted. Please refer to plate section for colour figure.

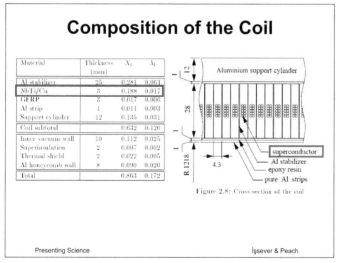

**Fig. 5.28** Example of an animated table, first part. Please refer to plate section for colour figure.

boxes. Finally Figs 5.28 and 5.29 show examples of the use of animation to step through a table, introducing the numbers in a well ordered manner so that the audience can easily follow the speaker.

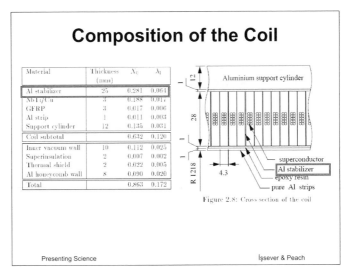

**Fig. 5.29** Example of an animated table, second part. Please refer to plate section for colour figure.

## 5.5 Animation

Animation can be a powerful tool to present a complicated dynamic process in a talk, or it can help guide the listener through a dense slide. In general, if you decide to use animation, its use should be carefully chosen, and not used too extensively. You should also make sure that you have a so called 'remote mouse' or 'presentation manager' when presenting your talk otherwise you will have to walk back to your laptop/computer to start each animation, which can be very clumsy and will disrupt the natural rhythm of your talk.

We have sometimes seen presentations where each of fifty lines of text is separately animated, which can become very tedious and tends to annoy your audience after a while. As with many things, less is more.

It can also be irritating to switch too frequently from one animation style to another, within a slide or between slides. For example, if you use an expanding style on one slide to introduce new objects, and a swirling style on the next, the audience becomes confused—what is the animation style trying to tell me? Is it telling me anything at all? So, keep it simple and consistent, unless of course there is a good reason to ignore this handy rule of thumb.

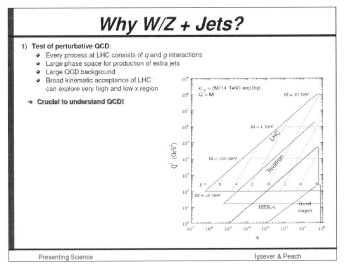

**Fig. 5.30** Example for an animation—slide 1. Please refer to plate section for colour figure.

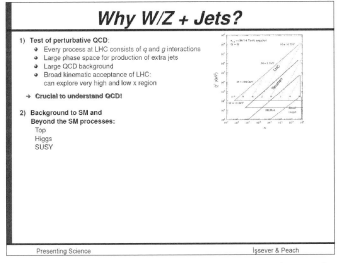

**Fig. 5.31** Example for an animation—slide 2. Please refer to plate section for colour figure.

There are several different types of animation. In-slide animation (revealing the contents of a slide in stages) can be very effective in making a point—see Figs 5.30–5.33. It can also be very irritating to the audience, a kind of striptease. Between-slide animation (slide transition) can be used to emphasize, highlight or add further information by duplicating the

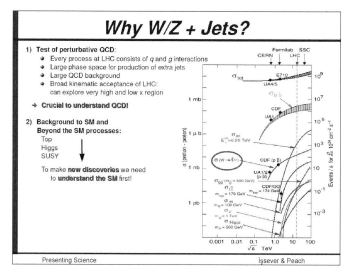

**Fig. 5.32** Example for an animation—slide 3. Please refer to plate section for colour figure.

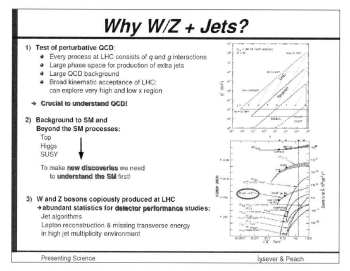

**Fig. 5.33** Example for an animation—slide 4. Please refer to plate section for colour figure.

slide, and making the two slightly different. Embedded clips (Quicktime, Realtime, video) can be very illuminating, **but beware—unless you can use your own laptop, it may not work as you expect when transported to a different computer**. You can use the timer to activate the next bit automatically, but again this is likely to be different

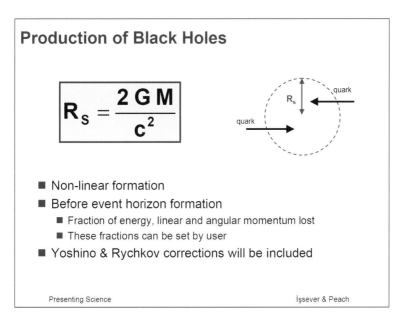

Fig. 5.1 Example of a slide for a working group meeting.

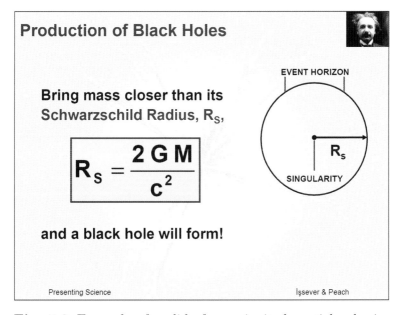

Fig. 5.2 Example of a slide for an invited particle physics seminar on the same topic as Figure 5.1.

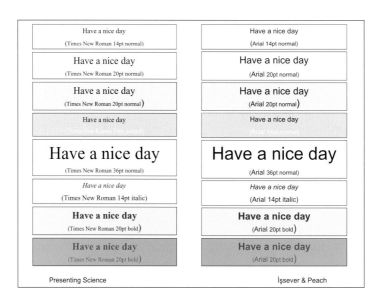

**Fig. 5.3** Different text font styles and background colours. This slide shows examples of serif (left column) and sans-serif (right column) fonts.

**Fig. 5.4** Different fancy font styles on different background colours.

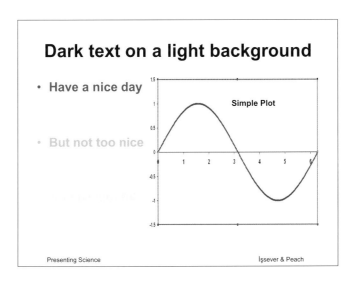

**Fig. 5.5** An example of a combination of a light–coloured background and different font colours. The green and the yellow font colours are not easily readable.

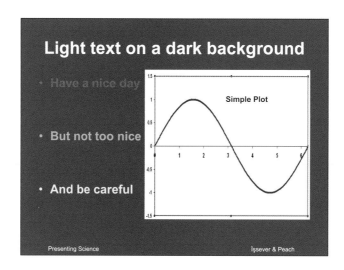

**Fig. 5.6** An example of a combination of a dark background colour and different font colours. The white and yellow fonts are clearly visible whereas the green font is less visible and the red font is very hard to read.

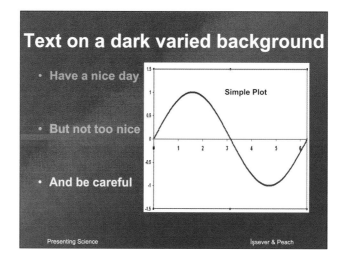

**Fig. 5.7** This figure shows a slide with varied background colours and different font colours. The white and yellow are readable but the red and green fonts are less visible.

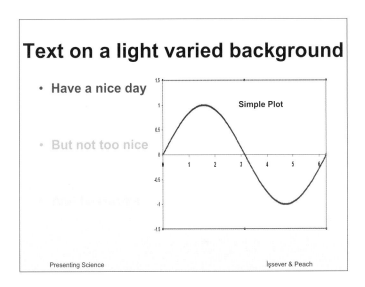

**Fig. 5.8** This is another example of a combination background and different coloured fonts. The yellow and green fonts are not a good choice whereas the red and black are readable.

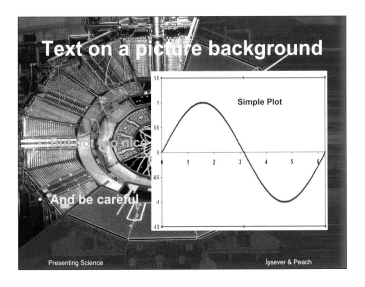

**Fig. 5.9** This is an example of a slide where a photograph was used as a background. The text on this slide is very hard to read.

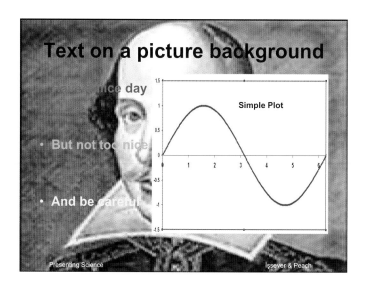

**Fig. 5.10** This is another example of a slide with a photograph as a background and again the features of the photograph are making it very hard to read the text on the slide.

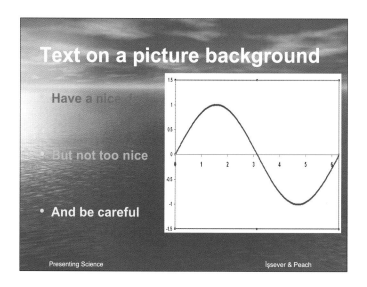

**Fig. 5.11** This slide uses a photo of a lake as a background which distracts from the text and makes it hard to read.

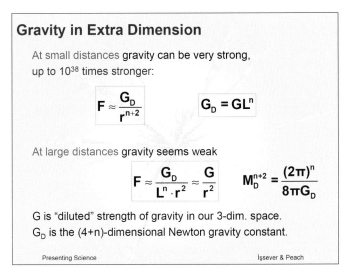

**Fig. 5.12** Slide showing a background picture which was modified such that it does not disturb the readability of the slide.

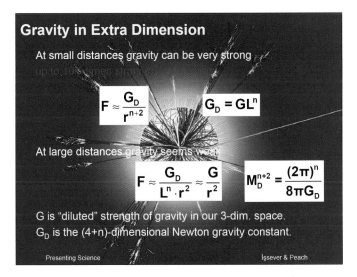

**Fig. 5.13** The same slide as in Fig. 5.12 where the background picture was not modified. Not all the font colours are working well with this background. This slide shows that one way to make text readable on a busy background is to use filled boxes around the displayed text.

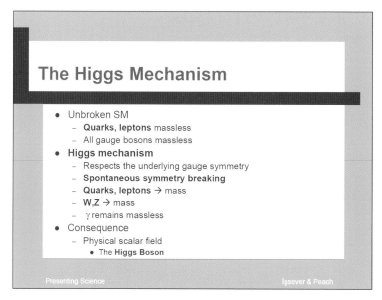

**Fig. 5.14** This is an example for a border style which uses a lot of space.

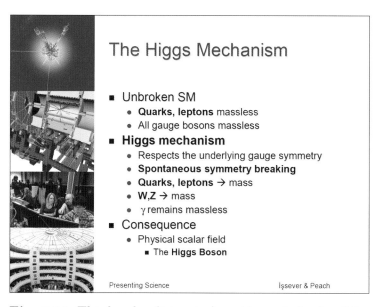

**Fig. 5.15** This border distracts from the main body of the slide and it also takes too much space.

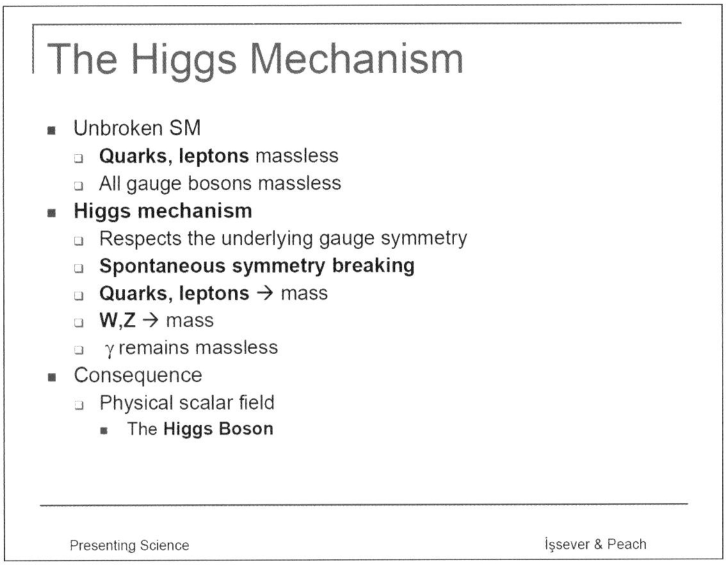

**Fig. 5.16** This is an example of a more economical border style where there is a lot of space for the main body of the slide.

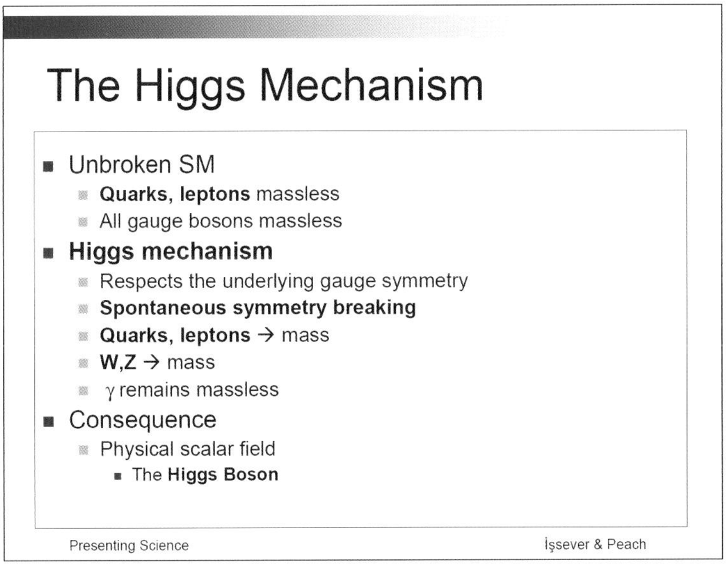

**Fig. 5.17** This shows another example of an economical border style.

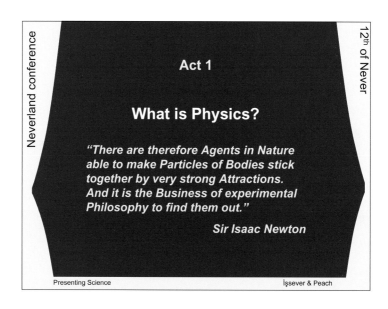

**Fig. 5.18** A final example of a combination of background and font colours and a 'theatre curtains' border.

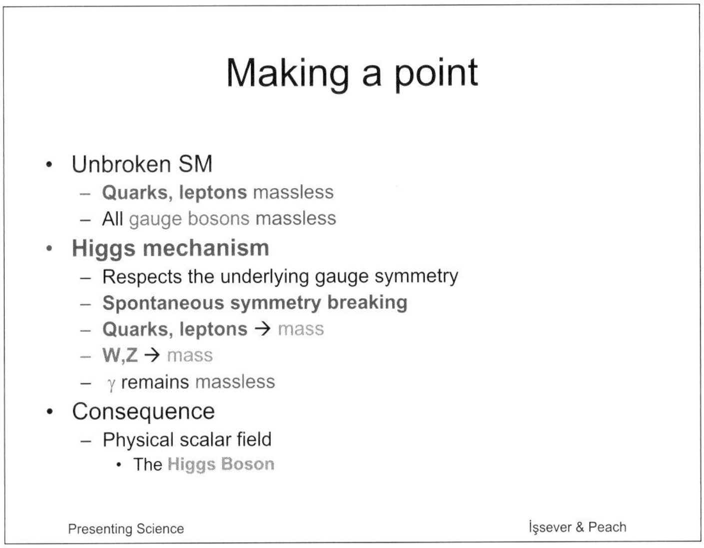

**Making a point**

- Unbroken SM
  - **Quarks, leptons** massless
  - All gauge bosons massless
- **Higgs mechanism**
  - Respects the underlying gauge symmetry
  - **Spontaneous symmetry breaking**
  - **Quarks, leptons** → mass
  - **W,Z** → mass
  - γ remains massless
- Consequence
  - Physical scalar field
    - The Higgs Boson

Presenting Science                                                    İşsever & Peach

**Fig. 5.24** In this example, the same information as in Fig. 5.23 is given added emphasis using colour. Blue and purple are used to distinguish between *fermions* and *bosons*, red is used to highlight the problem that the *Higgs mechanism* addresses (*masslessness*) and the main feature of the mechanism (*spontaneous symmetry breaking*) and green links the acquisition of *mass* by the fermions and bosons to the *Higgs Boson*.

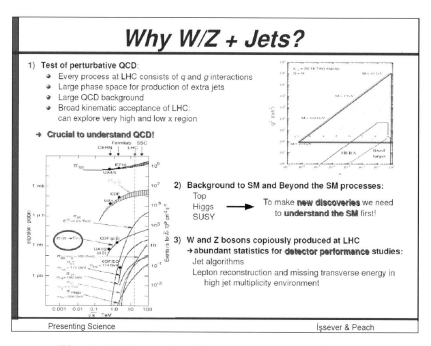

**Fig. 5.25** Example of how not to present figures.

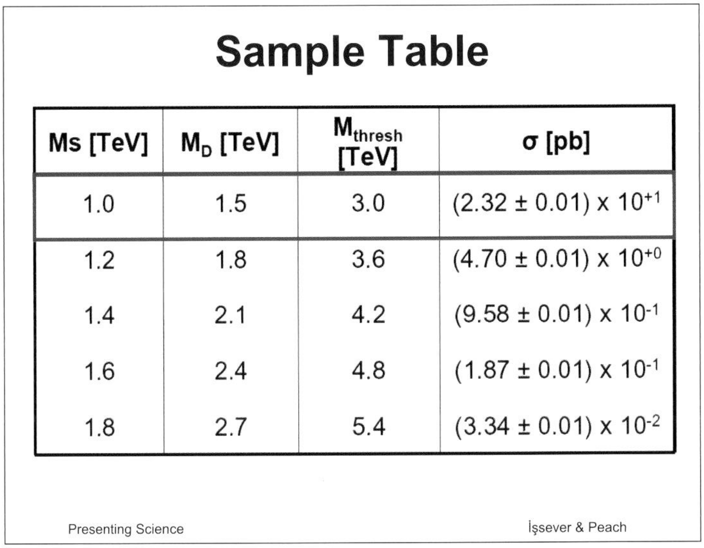

## Sample Table

| Ms [TeV] | M$_D$ [TeV] | M$_{thresh}$ [TeV] | σ [pb] |
|----------|-------------|--------------------|--------|
| 1.0 | 1.5 | 3.0 | $(2.32 \pm 0.01) \times 10^{+1}$ |
| 1.2 | 1.8 | 3.6 | $(4.70 \pm 0.01) \times 10^{+0}$ |
| 1.4 | 2.1 | 4.2 | $(9.58 \pm 0.01) \times 10^{-1}$ |
| 1.6 | 2.4 | 4.8 | $(1.87 \pm 0.01) \times 10^{-1}$ |
| 1.8 | 2.7 | 5.4 | $(3.34 \pm 0.01) \times 10^{-2}$ |

Presenting Science     İşsever & Peach

**Fig. 5.26** Example of a table which is not too dense, where a red box is used to highlight the row under discussion.

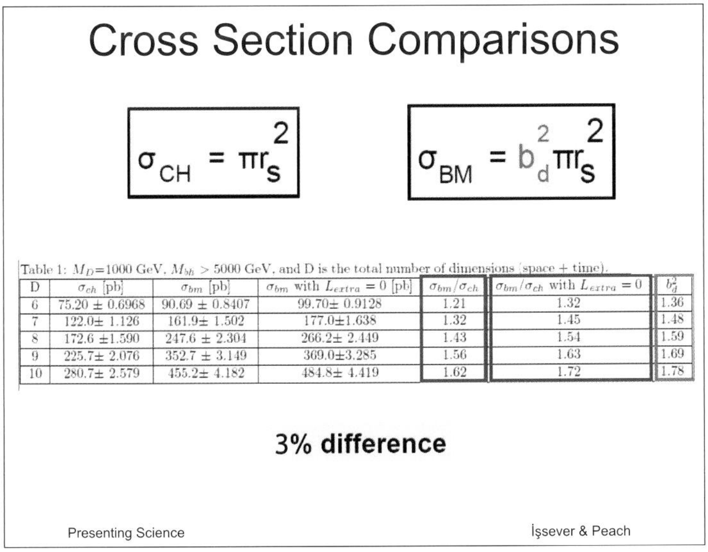

## Cross Section Comparisons

$$\sigma_{CH} = \pi r_s^2 \qquad \sigma_{BM} = b_d^2 \pi r_s^2$$

Table 1: $M_D = 1000$ GeV, $M_{bh} > 5000$ GeV, and D is the total number of dimensions (space + time).

| D | $\sigma_{ch}$ [pb] | $\sigma_{bm}$ [pb] | $\sigma_{bm}$ with $L_{extra} = 0$ [pb] | $\sigma_{bm}/\sigma_{ch}$ | $\sigma_{bm}/\sigma_{ch}$ with $L_{extra} = 0$ | $b_d^2$ |
|----|--------------------|--------------------|------------------------------------------|---------------------------|-------------------------------------------------|---------|
| 6 | 75.20 ± 0.6968 | 90.69 ± 0.8407 | 99.70± 0.9128 | 1.21 | 1.32 | 1.36 |
| 7 | 122.0± 1.126 | 161.9± 1.502 | 177.0±1.638 | 1.32 | 1.45 | 1.48 |
| 8 | 172.6 ±1.590 | 247.6 ± 2.304 | 266.2± 2.449 | 1.43 | 1.54 | 1.59 |
| 9 | 225.7± 2.076 | 352.7 ± 3.149 | 369.0±3.285 | 1.56 | 1.63 | 1.69 |
| 10 | 280.7± 2.579 | 455.2± 4.182 | 484.8± 4.419 | 1.62 | 1.72 | 1.78 |

**3% difference**

Presenting Science     İşsever & Peach

**Fig. 5.27** Example of a table taken from a publication, with important columns highlighted.

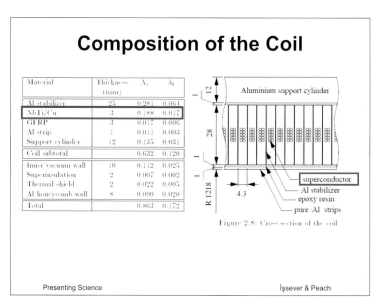

**Fig. 5.28** Example of an animated table, first part.

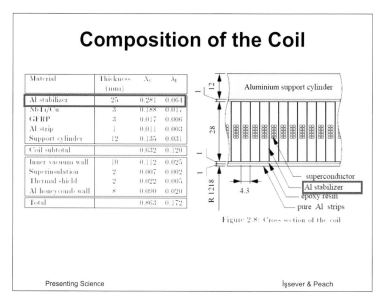

**Fig. 5.29** Example of an animated table, second part.

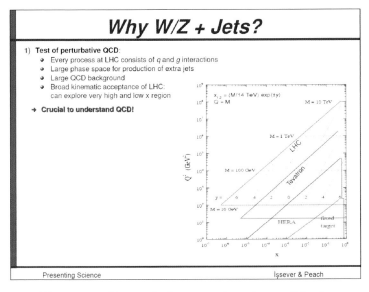

Fig. 5.30 Example for an animation—slide 1.

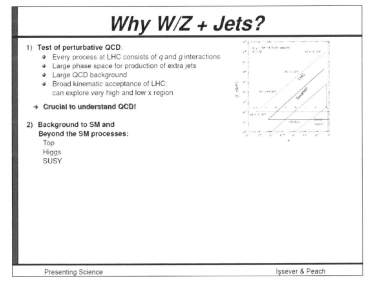

Fig. 5.31 Example for an animation—slide 2.

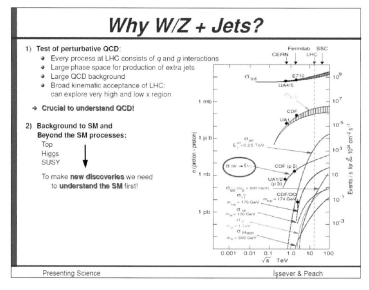

Fig. 5.32 Example for an animation—slide 3.

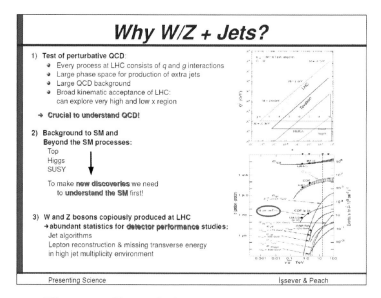

Fig. 5.33 Example for an animation—slide 4.

on different computers, and requires very careful timing of the words—it becomes more of a 'play' than a discourse. As with everything else, it is up to you. If you like it, and you think it helps the audience, do it, but do not overdo it.

## 5.6   Equations

*If a picture is worth a thousand words, a big equation probably requires a thousand words.* Of course, mathematics is the very essence of science, and equations are part of the language of mathematics. It is a very elegant, precise, comprehensive and compact way of expressing our scientific reasoning and arriving at our scientific conclusions. However, these features also make it a difficult language to read quickly, especially if the concepts or applications are unfamiliar. We therefore need to take special care in presenting equations in science.

The context of the talk has a big influence on the choice of what to show, and how to show it. What may be received as an excellent presentation in one context (say the professional seminar ) might be completely inappropriate in another context (say the public lecture ). Whatever the context, you should be prepared to spend a considerable amount of time talking the audience through the equation and slide.

Of course some equations are so well known that they have almost entered the language as entities in themselves—there can be few people who have not heard of $E = mc^2$. However, the fact that the equation is familiar does not necessarily mean that the equation is understood; whether you consider $E = mc^2$ to be right or wrong depends on whether you think mass is a constant or not. *So, if you show this simple formula in any context, you should probably take a moment to discuss how you* **interpret its meaning**.[2]

Before discussing how to present equations, let us make a general remark. We believe that you should always ask the following questions before including any formula or equation on a slide:

- *What do I expect my audience to learn from seeing this equation?*

---

[2]For those unfamiliar with the controversy, the issue revolves around the observation that $m^2c^4 = E^2 - p^2c^2$ in any frame of reference, so that in this definition mass is constant, independent of velocity. With a constant mass, the correct equation is either $E = \gamma mc^2$ (where $\gamma = 1/\sqrt{1 - (v/c)^2}$) or $E_0 = mc^2$, where the subscript 0 implies 'at rest'. The alternative interpretation for $E = mc^2$ is that $m = \gamma m_0 c^2$, which implies that mass varies with velocity, in contradiction to the spectral equation shown above.

- *Is there a better way of communicating the message to the audience, for example with a plot?*
- *How long will it take me to talk the audience through the equation?*
- *Do I need to show this equation at all?*

You are likely to come to the conclusion that, yes, you do need to show the equation, but the time will not have been wasted—you now know why you have to show it, and what you are going to say to help the audience understand it better, and you may have decided upon a helpful visual aid as well.

In the discussion below, we present some common situations and suggest ways of making the mathematics accessible. In each case, we will present the context, and discuss some options that might help you make your case effectively.

### 5.6.1   The formal derivation

Suppose that you have worked out a 'truly marvellous' proof of Fermat's last theorem, which the 'margin [of this book] is too narrow to contain',[3] but which is much shorter than the Wylie proof, and uses only concepts that would have been familiar to Fermat, so that it just *might* be what he had in mind. You can certainly expect to receive many invitations to present your work, and you will have to go through the proof in some detail. There is no alternative.

One option, of course, is just to display the paper that you have written on the proof page by page on the screen, but this might not be the most effective way of presenting the ideas—the number theorists will probably follow, but the merely curious might lose the thread half way through, which would be a pity.

An alternative would be to present the proof line by line, discussing each line in detail before passing on to the next, so that the audience can follow what you are doing, and have the time to absorb and perhaps question. In fact, look at the slides shown in Figs 5.34–5.37. As each new line of the proof is presented, the previous lines remain visible but faded, so that the audience can see where your current focus is, but can also refresh themselves about how you reached your present position. When the slide is filled, it is a good idea to repeat the last line of the previous slide, again faded, so that the audience can still follow the proof. You

---

[3]Fermat of course wrote in Latin: 'Cuius rei demonstrationem mirabilem sane detexi. Hanc marginis exiguitas non caperet.'

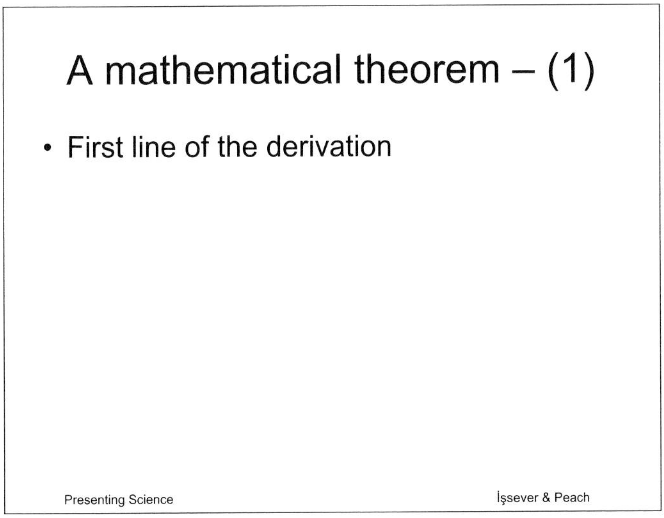

**Fig. 5.34** Revealing the derivation of a theorem in stages.

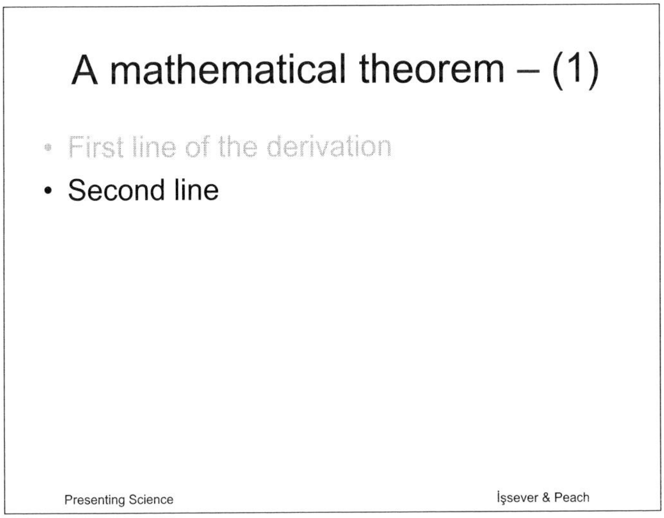

**Fig. 5.35** Revealing the derivation of a theorem in stages.

should explain carefully any hidden assumptions or suppressed steps in the argument; while it may be obvious to you, it may not be immediately obvious to all of your audience. There sometimes seems to be a reluctance to go into the details, perhaps because the speaker is afraid that the audience will feel offended by the fact that these 'obvious' steps are being

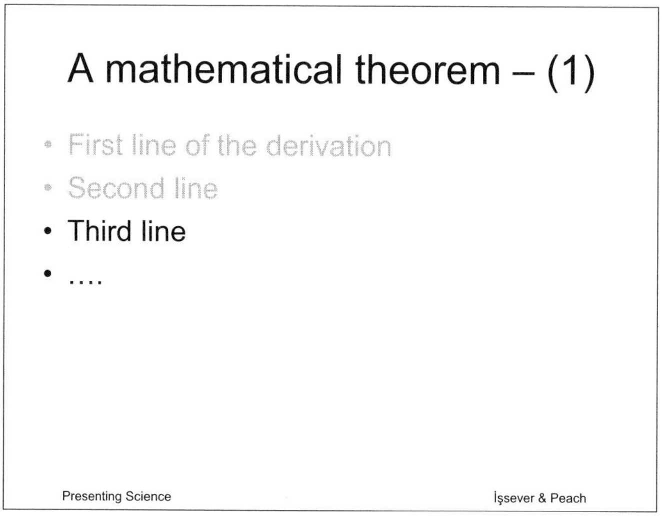

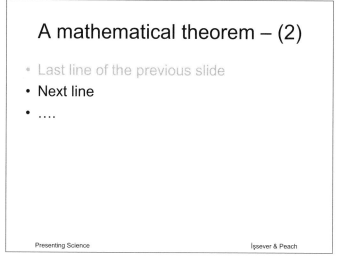

**Fig. 5.36** Revealing the derivation of a theorem in stages.

**Fig. 5.37** Revealing the derivation of a theorem in stages.

explained to them, or that by going into too much detail the speaker appears unsure about the ground. In our view, it is always better to explain too much for the sake of the graduate students than it is to risk losing most of the audience at a crucial point in the argument.

## 5.6.2 The professional seminar, conference presentation or group meeting

This is perhaps the most difficult context to judge correctly, because it can cover so many different situations. Let us start, then, with the two extremes.

1. The equation is well-known to the audience, who can be expected to be very familiar with it and its interpretation.

   Here, you do have to question why the equation is being shown at all, since it adds very little to the discussion. You may be showing it to demonstrate that you are using the right formalism, and do not need to spend much time discussing it, in which case you should indicate in some way that this is just a reminder, dismissing it in a few words 'For this analysis I used the well-known Somebody's equation shown here'.

2. The equation is unfamiliar or new, or is used in an unconventional way.

   Here you need to explain carefully what the equation means, otherwise you will lose most of the audience. Most people do not read mathematical equations quickly, and the notation might be very compact. $\Delta U = 0$ is a very simple equation, until you are told that '$U$ is the equation of state of the Universe'. While you may not need to go through the derivation in quite the same detail as in the previous example, you will need to explain what each of the terms in the equation mean, and draw the attention to the audience of its significance. If the equation is very long, or in a very compact notation, you should plan on spending several minutes discussing it.

Once you have decided what it is you must show, you then have to decide how to present it in a way that allows the audience to follow. Look at the equation shown in Fig. 5.38. Unless you are giving the talk to real experts in the field, you may be fairly certain that some in the audience will find it a little daunting. Talking them through it might take quite a long time. So, as with the formal derivation, the approach is to break the equation into a few steps, each of which introduces a new element into the argument, and discuss each as it is introduced. This example is worked through in detail in Appendix A.

While this looks to be a long way of introducing the formula, in fact it can be completed quite quickly for the professional audience. The point is that, at each step, you have explained carefully how it relates to the previ-

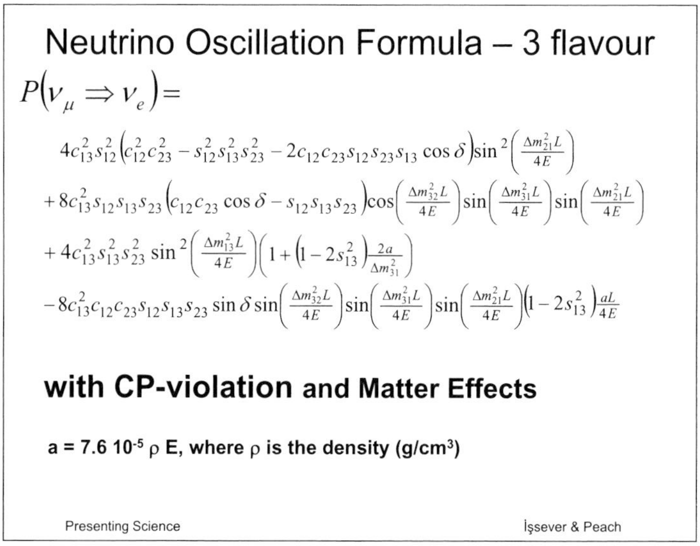

**Fig. 5.38** A complicated equation. It is in fact the full three-flavour neutrino oscillation formula, including CP-violation and matter effects. See Appendix A for a detailed walk through this slide and for a brief discussion of the physics behind the equation.

ous step, which they understood, and that each step introduces some additional complexity but also additional, and rich, phenomenology. Those who need to will work through the complexity later, once they have understood the phenomenology. And the graduate students will be pleased that they have at last understood it!

### 5.6.3 Public lectures and schools talks

In talks to the general public or to schools, you really do have to ask the question 'what is the purpose of showing this equation?' In most cases, it is likely that the purpose is better served by an illustration or some animation. In the case of neutrino oscillations, for example, a cartoon of a two-flavour oscillation and then a three-flavour oscillation evolving with time will convey the essence of the equation without the barrier of mathematics.

However, from time to time you may wish to remind the general audience that there is a rigorous mathematical derivation behind all of the pretty graphics, and showing them a complicated formula like that in Fig. 5.38 might be just the way to do it. There is no real chance of explaining its meaning in a public lecture, even with the animation shown in Appendix A. Nevertheless, if you do wish to show something like this to give a hint that the mathematics might be challenging, a useful approach might be the following. First, announce that you are going to show a rather terrifying equation. Then, advise the audience that there is no danger, but that if they are nervous in the presence of hard sums, perhaps they could hold hands with their neighbour. Finally, show them the equation rather briefly, remarking that it does look rather intimidating. After removing the equation and replacing it with some more easily accessible graphical representation of neutrino oscillations, make some casual remark like 'There, that was not too bad was it' and continue.

While this approach does not give the audience any real feeling for the intricacies of the mathematics, it does serve several purposes. It shows them that science is both rigorous and, at one level, difficult. Those in the audience who do understand some of the complexities are pleased that you have shared with them 'the real thing'. It may even stimulate some good questions at the end—'could you show that big formula again and explain why it shows neutrinos oscillating'. You may, of course, wish to go back to the simple two-flavour example (carefully included in the backup slides for just this eventuality) and quickly give some more detail. Further discussion can take place in the bar afterwards.

But again we return to the basic question—what is the purpose of showing the equation, and how do you expect the audience to respond?

## 5.7  Posters

These days, many conferences, large and small, have poster sessions. These are often places where early-stage researchers (graduate students and postdoctoral fellows) can show off their work. Some conferences have very large sessions, with perhaps hundreds of different posters being displayed each day of the conference. Other conferences may have only a small number of posters scattered around the coffee area. Whatever the arrangements, you will want to make sure that your poster is **attractive** and **informative**—if the delegates just pass by, you will not have the opportunity to explain the genius of your work, and with it the chance to make that lasting impression that might lead to your next job. Quite often, there are modest prizes for the best posters, which provide another useful entry on the CV.

Many of the skills that you develop for a visual presentation are also relevant for a poster. However, **everything has to fit on one** '**slide**'. You will need to know how much space is allocated. Convenient poster sizes are usually expressed following the ISO 216 standard, with A0 (841 mm × 1189 mm) or A1 (594 mm × 841 mm) being perhaps the most common. You may also need to know whether the space allocation is tall-and-thin (portrait) or short-and-fat (landscape); it rarely looks good if you have to bend your poster round a corner in order to make it fit.

*How many words.* Unlike the slide, the viewer has time to read, and re-read, the poster. The text can thus be arranged into sentences, and the normal rules of grammar respected. However, the viewer is usually just passing by, and probably does not have more than a minute or two to spare (there are many other posters to see, the queue for the coffee may have shortened, the next conference session may be about to start), and so it is not a good idea to have too many words. We have seen some 'poster displays' that consisted of about ten pages of typescript from a preprint stuck on the wall; we cannot recommend this approach, for all of its economy of effort—it looks awful and it is uncomfortable for the viewer to read. Ideally, the words and the pictures or plots should be **self-explanatory**; your job when standing by the poster is to add detail or commentary, and to answer questions. So, the number of words can be somewhat larger than you would use on a slide, but very many fewer than you would use in a written paper. Use a fairly **large font** (16 pt or even 20 pt for sentences, 30 pt or larger for titles and headings). Try to keep the paragraphs or blocks of text reasonably short (ten lines or fewer), with reasonably sized margins. Use bullets to emphasize key facts, and avoid too many tables.

*Pictures and plots.* The poster format just *loves* pictures and plots. Try to resist the temptation to put in too many pictures and plots; as with too many words, it is likely to deter the casual viewer, who will see the mass of information as hard work. So, choose the three or four most relevant *and attractive* pictures (preferably) or plots, and arrange them so that they draw the viewer into the poster. The eye can very rapidly identify patterns—make sure that the **flow of the poster** is natural, starting at the top with a clear trajectory to the key piece of information, not too near the bottom.

*Layout.* The layout of a poster and the layout of a slide are similar. The title should be clear and preferably quite short. The poster needs a 'frame', perhaps containing various logos, and a background. Your name and the names of your colleagues (if appropriate) will also need to appear somewhere. It is a good idea to lay out the poster as a series of 'blocks'—**title block**, **introduction**, **key message**, **further details**, **ancillary information**—and try to achieve a harmonious balance, with the emphasis on the key message containing the arresting picture.

*Final thoughts on posters.* A poster is basically advertising, in this case *advertising you and your work*. It must be attractive, easy to read, informative and accurate. The aim is to grab the viewer's attention—'this looks interesting, I wonder what it is about'. There is of course a tension between the various objectives set out above. Being attractive implies not having too much detail, which might conflict with the need to be accurate. Giving all of the information might conflict with making it easy to read. It is therefore a good idea to have readily available a **handout** or some further information—this is where your latest preprint comes in handy. Those who have have been enticed into conversation can be rewarded with a souvenir, something which does have all of the information, and the references, that they can read at leisure. Of course, human nature being what it is, much of this material will go unread too, but some who take it will read on and these may be sufficiently impressed to contact you afterwards. Think carefully about what you are going to say to the interested visitor. Have one or two additional facts ready to amplify and illustrate the importance of the work. Try to find out whether they are interested generally in the field, or whether your work might be *useful* to them, and respond accordingly. If it is possibly useful to them, make a note of who they are (swap business cards if possible).

Finally, always be enthusiastic about your work.

## 5.8   Visual aids and props

It is sometimes helpful to have some visual aids or props to illustrate a talk. These can take many forms—a piece of the equipment used in the experiment, some leaflets explaining the background, a demonstration experiment (if it is a public or school lecture, preferably ending with a small explosion). Such aids can enliven any presentation. However, they also take time to introduce and present. As so often, the context is important in deciding whether it is worth the bother, and the nature of the visual aid will vary dramatically with the context.

For the professional occasion, the visual aid needs to be really relevant, otherwise you risk antagonizing your audience or your colleagues.

Nevertheless, there are occasions where it is appropriate—for example, if you have just received the first production module from the manufacturer.

As a rough rule of thumb, the more 'public' the occasion, the more useful a visual aid or demonstration is likely to be. The annual Royal Institution Christmas Lectures, which Michael Faraday delivered nineteen times between 1820 and 1860 and which are now televised, provide a very good example of the excellent use of demonstrations and visual aids. Of course, the involvement of television in these lectures means that there are significant resources available to devise the demonstrations, and talented staff to implement them. But the requirement to make 'good television' means that the demonstrations and the accompanying need to involve members of the young audience can intrude in the development of the argument.

If you have decided to use a visual aid, then there are some sensible precautions that you need to take.

- Practise how you introduce it—you want it to achieve the maximum impact (otherwise, why are you bothering) and so you want to be sure that the audience understand its significance.
- Allow sufficient time for the audience to appreciate the purpose of the display.
- What happens after the demonstration—how do you get the audience to concentrate their focus back onto *you*? You can be fairly sure that, if your demonstration uses a live rabbit, most eyes will stay on the rabbit while it remains visible on the podium.

Try to avoid anything too complicated—the audience (especially its younger members) are unlikely to be impressed by a statement like 'well, it worked this morning'. On the other hand, the passive display of a small piece of moon rock, barely visible from the front row, will not impress either.

So, if you are going to use a visual aid, think carefully about its purpose and practise thoroughly its delivery. If you do it well, you may receive a round of applause, and quite probably a request at the end to 'do it again'.

## 5.9 Summary

The style of the presentation or poster is important. It says quite a lot about you. All of the features of a slide—the font, the frame, the colours, the use of pictures, plots and tables, animation, equations and visual aids—say something about you, and should be chosen with care. As so often, if you make the right choices (that is, those consistent with your personality and your message) it will pass unnoticed—you will have given a good talk; but make the *wrong* choices, and your audience will be disappointed, and are likely to withhold the ultimate accolade.

# 6
# Preparation and presentation

*Semper ad eventum festinas et in media res*
*Non secus ac notas auditorem rapit*[1]
*Horace (55 BC–8 BC), Ars Poetica*

Horace was clearly being critical. We need to ensure that we take our audience with us on the journey. In this chapter, we will discuss some of the ways in which you can try to achieve that happy outcome, once you have mastered your brief and mustered your material.

Presenting scientific results is a serious business, but this does not mean that it has to be dull and boring. Of course, a lot depends upon personality; not everyone is outgoing and extrovert, or witty and erudite, and it can be quite difficult to follow someone who is. Nevertheless, like many things, it is a skill that can be acquired with practice. The point is that presenting scientific results is one of your professional duties, and it should be done professionally.

It is likely to be some time in the development of your career before you are invited to give a major address to a large audience at a major event, which is when all of your oratorical and theatrical skills will be needed, but it is a good idea to start to practise them sooner rather than later. There is a trend towards giving younger scientists an opportunity to present their work at a major conference—if you are offered the chance, take it, but prepare well. In any case, many wise group leaders know that an enthusiastic graduate student or postdoc always makes a very good impression on visiting committees and suchlike, and so you might find yourself giving a keynote presentation quite early in your career.

This is where self-awareness is vital. **You need to assess your own strengths and weaknesses**. Are you extroverted or introverted? Are you confident or timid? Is your voice strong, projecting well to the back of the hall, or do you speak with quiet authority? Are your nerves steady, or do

[1]He always hurries to the main event and whisks his audience into the middle of things as though they knew already.

you perspire and shake? In reality, most of us are somewhere in the middle most of the time. Even an experienced communicator is likely to be a little nervous if there are several Nobel Laureates in the audience. Then there is the fear of **stagefright**—what would happen if I were to forget a key part of the argument, or if I could not remember my collaborator's name, or made some other elementary error? It rarely happens, but this does not make the fear of it any less real. And, in fact, this fear is probably a good thing—it raises the adrenaline levels and prepares you for the struggle ahead. If you lose this fear, you have probably lost interest in the subject, and do not really care whether the audience find it interesting or not. In that case, it is time to find a new subject to talk about, and write a new seminar!

We are all different, and react differently in different situations. In the sections below, we offer general advice, and reveal some of the tricks that we have used, or seen used by others, to make an effective presentation. As you develop in your career, moving from 10-minute presentations of work in progress to your immediate group, through short conference presentations to major conference events, **practise your skills**. Of course, you need to be able to control the 'volume' of your chosen techniques to suit the occasion, but this is also part of your professional development.

Now, before we address a few technical issues, we should make one thing quite clear. The most important thing that you must do before standing up and speaking is to have fully prepared the talk. As we have stressed several times, you are responsible for everything on each slide, and also responsible for everything that could have been on each slide, but is not. If challenged, you need to be ready with an answer—why did you include this, why did you exclude that? While this sounds daunting, and if you are about to address a major international conference, it is daunting, it should not be *if* you have prepared the talk carefully, and you are on top of your subject. At smaller conferences and shorter talks, **keep strictly to the brief**—in general it is better to cover rather less ground well than to present an overview of the entire field superficially. These small conferences are your training grounds. Accepting an invitation to give a seminar in some distant university can sometimes seem like a chore—a whole day spent on the train when there are results to analyse, there is apparatus to build, and there are papers to write. But our advice is 'always accept if you can do it'. The more you practise your presentational skills, the more proficient you will become. Remember the words of the great South African golfer Gary Player[2] 'the more I practise, the luckier I get'.

However, you should also realize that there are two things on your side. Firstly, if you are standing at the podium with the microphone, then you are (or should be) in charge; you are the authority figure. Secondly, the audience is usually friendly, willing you to give them a good talk; they don't want to feel that they have wasted their time.

Here we need to make another side remark. This assumption that the audience is generally on your side is, in our experience, a good one for scientific presentations. However, in other spheres—politics for example—this may not be a very good assumption, and you would need to prepare yourself very differently by, for example, trying to anticipate the strong objections and rehearsing with good friends how to deal with frequent hostile interruptions or aggressive questions. Unless your subject is very controversial—say, arguing for intelligent design or any of the 'bell-curve' theories of intrinsic intelligence—the audience is likely to listen with polite, even rapt, attention. If the audience is small, you may be interrupted with polite questions or requests for clarification, but you are unlikely to encounter seriously disruptive behaviour. The worst that is likely to happen is that some people might leave before the end—perhaps they have a train to catch, or their mobile phone has vibrated, or they have realized that they are in the wrong talk, or maybe you have failed to entertain them enough. However, they are unlikely to make a fuss.

So, take the advice given below as it is intended—some hints about how to create a good impression. But, as we have emphasized several times, develop your own style. If it works for you, go with it.

## 6.1 Position to audience

In many circumstances, your position with respect to the audience is determined by the layout of the room or theatre in which the talk takes place. Nevertheless, there are still choices that you can, and should, make. First and foremost, you should try to **face the audience** as much as possible. This signals to the audience that you are confident in your results. However, this is where modern computer technology creates a barrier, not present with the previous 'advanced technology' (the overhead projector).

---

[2]There is some debate about who actually said it, but it is generally attributed to Gary Player. Samuel Goldwyn said something similar—'the harder I work, the luckier I get'.

With foils and an overhead projector, it was possible to face the audience most of the time, turning away to the screen only occasionally. It was possible to indicate the part of the foil that you were discussing by pointing to the transparency—the grubby finger or tip of the pen being also projected behind you. With a computer presentation, you are more-or-less obliged to use a laser pointer, especially in a large lecture theatre, and this means turning away from the audience and towards the screen. This was also a problem with the earlier 'chalk-and-talk' technology, where all too often the lecture was delivered to the blackboard. It is possible, with practice, to use the computer-generated mouse to indicate which part of the slide you are talking about, but the moving arrow is rarely very visible, and to make it visible, there is a tendency to wave it around rather alarmingly, which is distracting. Simple animation and use of a remote mouse can be very effective. For example, highlighting or circling the point currently under discussion allows the audience to see clearly where you are on the slide, while you can continue to face the audience, maintaining contact with them; you can check that what is being displayed is what you expect by glancing at the laptop or monitor. If you do use the laser pointer to indicate which part of the slide you are now discussing, do so quickly and efficiently, and turn back to the audience for the discussion.

The **layout of the room** or theatre will determine whether you stand to the left or the right—if there is a projection of the talk, it is rare that you stand in the middle. This means that it is almost impossible for both you and your slide to be simultaneously the focus of attention of the audience. To minimize their discomfort, you should increase your discomfort by reducing the angle between you and the projection of your slide to the minimum. You can usually see the slide displayed on the podium monitor, so you can be confident that they are seeing what you are seeing, but you may not have a good view of the actual projection, especially in a large theatre. In smaller venues, for example the typical seminar room for twenty people, you have more choice. Note also that, in small venues, some people who have difficulty in hearing appreciate being able to see your lips move—this helps them enormously.

You can also use your position to the audience to signal changes of pace or the end of a section. For example, if there is a lectern, for most of the talk you might be behind it giving the main part of the talk. However, when you wish to give the audience a short break, by giving them some additional information (a story or anecdote, or a commentary), move away from the lectern. When you have finished this little aside, return to the lectern and continue to the next topic. If there is no lectern, you could adopt a slightly different pose—deliberately place the remote mouse or the laser pointer on the table while you digress, and then pick it up when you wish to resume.

## 6.2  Voice and language

The audience are not going to learn very much if they cannot hear you clearly, or if they are unable to understand what you are saying. A microphone will, of course, amplify what you say so that those sitting at the back can hear you, but it will not help if your speech is not clear. We will deal with some of the issues concerning microphones below. In this discussion, we will assume that you are addressing a medium-sized international audience of about 100.

The aim should be to speak in a **steady voice**, with the words clearly enunciated and the sentences clearly delineated. Even with a microphone, the level of the voice should be raised somewhat (but well below a shout), to impart confidence and authority. Try to avoid too many 'ums' and 'ers', or annoying fillers such as 'OK' or 'right' as you tick off the points that you want to make. This is where the rehearsal comes into its own. You need to make sure that your brain, tongue and teeth have grown accustomed to working together on the words and phrases, so that they flow freely. This is especially important if there are any difficult words or phrases—how do you pronounce 'methylidynetris-aniline' or 'Przewlocka'?

In general, and especially in an international environment, the language should be kept simple and straightforward, avoiding jargon, in-jokes, unexplained technical terms, acronyms and abbreviations. *It is especially important that native English speakers take care to address the needs of those in the audience for whom English is not the first language.* Sentences should be kept short, with few clauses and fewer strings of adjectives. Speak slowly and clearly, avoiding dialect words, slang expressions and regional pronunciations.

## 6.3  Microphones

Where the audience is larger than thirty or so, it is quite likely that there will be a public address system, so that those at the back can hear. The microphone can pose a surprising variety of problems. A well-adjusted public address system often means that you do not receive a great deal of feedback, of either the good kind (a prompt amplified version of your voice) or the bad kind (the piercing whistle), and so it can sometimes feel that you are talking to yourself.

There are several types of microphone. In our view, easily the best (but sadly, the one least often encountered) is the *head mounted radio microphone.* The distance between the microphone itself and the mouth is constant, so that the voice modulation is more natural, and the one that you intend. Because it is a radio microphone, it is also possible to move freely around the stage. The only real downside is the slightly self-conscious thought that you look a bit like a puppet from *Thunderbirds,* but this soon passes.

The next most convenient is the normal clip-on or *lapel radio micro-phone.* While these are very convenient, there are one or two things to keep in mind when using them. Firstly, if you suspect that there is going to be a clip-on microphone, try to dress accordingly. Tee-shirts and woolly jumpers do not work well, because there is almost nowhere sensible to clip the microphone—the neckline is too high and too close to the throat, and the clip is rarely strong enough to stay in place if attached to the front. It is also necessary to position the microphone carefully, biased slightly to the side between you and the screen. If you bias on the opposite side, then you become inaudible every time that you turn towards the screen to emphasize a point, with the result that the audience know that you said something important, but have no idea what it was. Secondly, there is always a radio transmitter to be parked somewhere. The most conve-nient is to use the pocket of a loose fitting jacket; the least convenient is a pair of very tight-fitting fashion jeans. Occasionally, the roving micro-phone might have a cable replacing the radio link—there is a very good chance that you will trip over the cable and rip the microphone from your clothing. This will provide an amusing interlude for those who like slapstick comedy, but is generally disruptive when it is not dangerous. Health and Safety considerations are gradually replacing these with radio microphones.

In some conference centres, often those in hotels, there is a *fixed podium microphone.* This poses some problems. In order to have a natural voice modulation, the distance from the mouth to the microphone should not vary too much, but this means that you are more-or-less rooted to the spot. This is usually acceptable, even preferred, for a 'political' speech, but is rather restrictive for the usual scientific presentation. If your style is to move around a lot during your presentation, then you will feel un-comfortable, but there is no alternative—grip the side of the lectern and stay in one position. You will need to use voice modulation and hand gestures to break up the content into manageable chunks.

In some lecture halls the microphone is *hand-held*, sometimes a radio microphone, often attached by cable. These are a bit of a nightmare, frankly. You lose one of your hands because you have to hold the microphone, leaving you to juggle the laptop and the laser pointer with the other. The only real advice is to keep the microphone hand steady (so that the distance between microphone and mouth does not vary too much) and keep reasonably still. However, you will very likely feel uncomfortable whatever happens, and will be relieved when it is all over.

Finally, when giving a presentation over the telephone or video conference, it is very difficult to judge the volume correctly, but try to keep a reasonable distance between your mouth and the microphone to avoid too much 'heavy breathing' being transmitted.

## 6.4   Using the laser pointer

The laser pointer is a very simple device—a red or green dot is displayed on the screen to enable you to highlight a particular feature on the slide by waving your hand. Given that it is so simple, it is remarkable how often the laser pointer is poorly used. The point of a laser pointer is to point at something on the screen. *Occasionally* it might be useful to describe a curve on the screen, for example if you are explaining the workings of a cyclotron, but in general **the pointer should be held steady**. It is rarely a good idea to wave the pointer wildly all over the screen to indicate the rough area that you are currently discussing; the eye and brain are programmed to follow movement, and rapid movement distracts from the slide content. If you doubt this, borrow a large lecture theatre, display a complicated slide, sit near the front and ask a friend to wave the laser pointer in ever larger circles rapidly over a part of the slide. How long does it take before you feel like throwing up?

Of course, if we are a little nervous, our hand may shake a little. However, tempting as it is to disguise this by waving the pointer around, resist. As with much else, try practising in an empty auditorium so that your hand–eye coordinations becomes natural—familiarity should reduce the shake. If there is still too much shake and wobble, then learn how to get the pointer directed at the right area of the screen, switch it on and move to the place on the slide that you wish to emphasize, underline or draw a box round it quite deliberately but not too slowly, and then switch off the laser pointer. The audience will now be focussed on the right area and can listen to you again.

Many venues these days have a simple **presentation management device**, which has 'next slide' and 'previous slide' buttons and an integrated laser pointer. If you have the chance, practise a little before the talk just to train your fingers and thumb so that you don't move in the wrong direction, or go on to the next slide when you meant to switch on the laser pointer. But some remote presentation managers have many more buttons, some of which do alarming things like exiting the presentation, starting again or skipping to the end. Here you really should try to play with the device for a few minutes before the talk. While the first time that you press the wrong button may be amusing, by the fifth or sixth time the joke will have worn rather thin, especially if you are the fourth or fifth speaker in succession to have the same difficulty.

Remember too that it is a *laser* pointer and not a stick, so **turn it on** when you point to something on the slide. The laser pointer is too short to be seen clearly by the audience (even in quite a small venue) and so they have no real way of seeing what you are highlighting. There is then a tendency to move closer to the screen, waving hands and arms about rather alarmingly while obscuring part of the slide if you have not turned it on. If you feel more comfortable with a stick, check with the organizer before the talk and make sure that one is available.

Always remember that it is a laser pointer. While many are low power, no-one in the audience is going to appreciate having it directed at them, so make sure that you do *not* point it at the audience. Some venues will have restrictions on the power of the laser pointer that you can use—if you wish to use your own laser pointer, check with the management beforehand if you are unsure.

As a final comment, insisting on mastering the laser pointer may seem a bit fussy—audiences are likely to be tolerant of a bit of shake, the odd mistake and some waving around. But this is beside the point. Giving a 'good talk' means trying to make optimal use of all of the technology available to help the audience. Your excellent pointer technique will be much appreciated, especially if you follow a speaker who has spent the previous half-hour drawing Lissajoux figures all over the screen to no great purpose.

## 6.5 Dress code

You may think that what you wear has no influence on how your talk is received, and often you will be right—the message is more important than the messenger. But as with everything else, **the way you dress should be consistent with your personality**. If you dress strikingly it can help you be remembered. Some people have difficulty in remembering names, but perhaps have no difficulty in remembering the bright red bandanna and the purple-and-yellow checked shirt or blouse. Our best advice is 'try to be slightly more smartly dressed than the audience'. If you turn up for the regular 9 a.m. group meeting in full evening dress, people may conclude, probably accurately, that you have not had much sleep. Wearing a tee-shirt and jeans when the audience is in evening dress is certainly making a statement, but it may not be a statement that the audience wishes to hear, and they are then quite likely to dismiss the other statements as well.

## 6.6 Entertaining the audience—jokes and asides

As we have repeatedly emphasized, the scientific presentation should inform, educate and entertain, in some proportion. While entertainment may not be the main motivation, we believe that you do have a responsibility to your audience to make the experience a pleasure as well as the duty that it sometimes is. This is particularly true for the more public talks—popular lectures, keynote addresses, talks to schools or pieces to camera. Humour provides a kind of 'punctuation mark' in speech, defining the end of a thought and giving the audience a 'hook' which will help them remember the point you were making.

But humour is a difficult subject, and its acceptability depends upon many factors. Failed attempts at humour can be deeply embarrassing. Some remarks can offend, even if they are not intended to be offensive. Light-hearted remarks about one's hosts that may be acceptable in one culture can be completely unacceptable in another.

*A simple example.* A noticeboard might be insecurely attached and fall off during your talk, making a loud bang. Depending upon the circumstances, it might be considered very funny or outrageously offensive to remark that 'this place will be really nice when it is finished!' So, the first piece of advice is 'if you are not sure, play it safe and carry on with the science'.

Humour is also a very powerful communication tool, and sharing a joke can create a bond. This is particularly so if the joke works on several levels—if it is not only funny for anyone, but also has a special meaning within the context of the talk. But try to avoid anything too complicated, especially in front of an international audience. If only a minority of the audience see the joke, the majority feel excluded.

Of course, delivering a scientific talk is not the same as doing stand-up comedy; for one thing, the audience is likely to be more sober for the former than the latter. Nevertheless, it is possible to learn a great deal about public speaking by watching a good stand-up comedian work a crowd. This is not the place for a treatise on comedy, but a few observations might be appropriate. The comedian usually adopts a particular persona, and derives the comedy by acting either *within* the character if the character itself is comic or *against* the character if the character is pompous or serious. The most important point is that the words are very carefully chosen, and the timing is carefully rehearsed. Even *ad lib* comments are carefully planned. Finally, whatever the persona, the stand-up comedian has to be self-confident.

If we now turn to the scientific presentation, what can we learn from this? Firstly, the aim should be to act always within the appropriate persona—the serious scientist. Humorous remarks should therefore be derived from the work, and not be disconnected from it. For example, an 'Englishman, Irishman and Scotsman' joke is rarely likely to be relevant, even if it is not offensive. However, if you are the 'Cholmondely' of work by 'Cholmondely, O'Reilly and McPherson' you might get away with a remark like 'we sound like an Englishman, Irishman and Scotsman joke'.

If you do want to be humorous, then as you prepare your talk, you should be thinking in the background 'is there any **anecdote** or **amusing aside** that can be included here'? Any such asides or remarks should be very brief—again learn from the stand-up comedian. The key thing here is timing, and to make it sound spontaneous, as if the thought has just occurred to you. This takes practice, firstly to get the words right and then to get the timing right, with the pause in the right place and of the right length.

*While the aim to is appear spontaneous, it can be very dangerous actually to be spontaneous.* Humour lies deep in the subconscious, and if it is allowed to emerge unfiltered it can lead to embarrassment or genuine offence. Suppose, for example, that you have just made a witty and erudite comparison between your eight key parameters, a musical octet and the Buddhist eightfold way. It suddenly occurs to you that you can round this off by saying something like 'adding a ninth parameter would be as strange as the sound of one hand clapping', neatly tying the musical analogy with Zen Buddhism. Before you have finished speaking, you realize, just before the audience falls silent, that your host has just recovered from the amputation of his left arm! Your subconscious has betrayed you (again?). Over the lunch, it noted the amputation, and merged your carefully rehearsed comments with the observed awkwardness with the knife and fork, and your subconscious made the disastrous connection.

Of course, visual humour (on the slides) cannot be presented as spontaneous, but it can be very effective and can surprise. There are a number of relatively simple techniques that can enliven any talk. However, as always, the visual joke must have an identifiable connection with the subject.

- A quite popular trick is to use an extract from one of the more intellectual cartoon strips—Dilbert, Doonesbury, Peanuts, etc.—which makes a wry comment on the subject of your talk. This can be very effective because the strip is itself funny and the context of the talk and its relevance makes it doubly so. However, finding the right strip can be difficult and so it helps if the strip is a part of your normal reading matter, and you take note of useful material. Be careful, however, of the copyright laws; you will probably get away with it in the oral presentation but there might be problems if it appears in the version that is posted on the web, and you really do need permission if it appears in the written version of the talk.

- An appropriate quotation at the end of the talk can underline your key message. Again, it is better if the quotation appears as something that might be part of your normal reading matter. There are, of course, many ways in which you can trace a quotation that you vaguely know but cannot quite remember the source, and this can sometimes reveal additional information that might be interesting to the audience.

- Suitably poignant photographs (the failed detector poking forlornly out of the wastebasket) can often express sentiments eloquently that would otherwise be difficult to express, eliciting both sympathy and smiles. It can also be informative to compare the 'artist's impression' of the experiment from the proposal with reality. However, 'holiday snaps' can be rather dull, although a picture of the barbecue held immediately after the experiment really worked for the first time adds a nice human touch.

Humour is a risky business. Some in the audience may feel that science is too serious to be treated lightly, and others (for a variety of reasons) may fail to see the joke. What may be acceptable in some cultures (for example jokes about haemorrhoids) may be unacceptable in others. Jokes about people, either present or absent, should generally be avoided. Likewise, try to avoid making self-deprecating remarks, or making remarks that undermine the seriousness of the work that you are presenting.

## 6.7   Are you sitting comfortably? Dealing with nervousness

We all feel nervous before a talk, even if it is just a short report to the group. If it is a big occasion, we are quite likely to feel very nervous. Do I have enough material, or is it far too much? Have I forgotten to include something vitally important, or has an unfortunate error crept in somewhere? Will my attempts at humour be well received, or will there be an embarrassed silence? How do I look? How do I sound?

Up to a point, this is very healthy. The body and the mind are preparing themselves for the task ahead. There are several 'Do's' and one big 'Don't' that can help.

- **Do** make sure that you have prepared the talk well. Refresh your memory about 5 or 10 minutes before your starting time.
- **Do** make sure that you are comfortable, whatever Enoch Powell might have thought about the benefits of a half-full bladder.
- **Do** check the auditorium, the podium, the audio-visual equipment well before it is your time to give the talk. If yours is the only talk (for example, the department seminar), this will only take a few minutes. If this is a conference talk, attend a previous session and then look around the podium, talk to the organizers, check out the equipment. If yours is then opening talk, then make sure that you get there *at least* 20 minutes early, and possibly a little more. The conference organizers will be pleased to see their first speaker (that is one potential disaster avoided) but they will still probably be trying to get to grips with the audio-visual equipment.
- **Do** have a bottle of water available, or check that the organizers have water available, to deal with 'dry throat' syndrome, or the tricky cough.

- **Do** try to relax your muscles by stretching a little.
  but

- **Don't** take any stimulants (alcohol or drugs). While these often can reduce inhibitions and make you feel more relaxed, they also reduce our ability to judge sensibly. Adrenaline is the body's natural stimulant—learn to trust it. Of course, once the talk is over, you may appreciate a large drink (and if it has been a good talk, several people may offer to buy you one) and that is just fine.

The first and most obvious help in tackling nervousness is to be absolutely confident of your material. This means starting to **prepare the talk in good time** (for a major event, several weeks ahead of time). In general, it is better to try to cover a limited domain in depth than to try to cover everything superficially—you are fighting (the adrenaline) on your own territory. As your self-confidence and reputation grow, you can broaden your domain, where your new authority (the authority that comes from your confidence and the authority that comes from your acknowledged expertise) will stand you in good stead.

The second important way of combating nervousness is to **rehearse**, alone or with a few good friends. Speak the words out loud, so that your brain, mouth and tongue are working together instinctively. This is particularly important if you are giving a talk in a second language.

The third way of dealing with nervousness is to have a 'cheat sheet' to hand. This can take many forms, and is rarely needed, but its existence (perhaps known only to you) gives comfort. There are three basic 'cheat sheets' that might come in useful.

1. A printout of the slides, one or two per page, upon which you have written in big bold pen the key point of each slide. Alternatively, you can use something like the PowerPoint notes option to add a text aide-memoire to each slide. It is probably better to use a few bullets than to write out verbatim what you intend to say, for two reasons. Firstly, if you do get lost on a slide, you need to find your place very quickly. Secondly, if you do write it down verbatim, there is an inevitable tendency to simply read what is written, and this leads to a rather dull and unconvincing delivery.

2. Have a printout at nine, sixteen or twenty-five slides to a page. This works well if your slides are very clear and uncluttered, so that you can see at a glance where you are and what the next slide contains. This is also very useful if you have to skip a few slides because time

is running short—in PowerPoint just type in the number of the slide that you want to jump to and type return.

3. Have a sheaf of notes with some big banners containing your key points and a few notes.

In fact, this can be seen as being professional about the talk, showing that you have given great thought to what you want to say, and how you want to say it. If you are nervous, or even if you are not, having the notes clearly visible on the table or lectern and consulting them *very* briefly every few minutes can be quite effective—you are just checking that you haven't missed anything out, and the brevity of the consultation confirms your have done you job professionally and are well on top of the subject. Then, if you do lose your way, a slightly less brief consultation should get you back on track.

**We strongly advise against reading from notes or from the screen word for word, however nervous you might be.** There are two problems with reading from notes; firstly, the delivery is likely to be unnatural and thus less convincing, and secondly if you really *are* nervous, the paper is likely to shake visibly. There are similarly two problems with reading from the screen; firstly, the audience can read it just as well for themselves and very likely will do so, and secondly you will have minimal eye-contact with your audience, again adding to the impression that you are not on top of the subject.

Note that there is at least one exception to this advice, underlining our view that in the end there are no absolute rules governing presentation. It sometimes happens that a keynote speaker is unable to attend the conference, but that he or she has sent the slides and if you are lucky, some speaker's notes. Rather than cancel the talk, the organizers have asked you if you can step in at the last minute and give the talk instead. Under these circumstances, the audience will understand why you have not prepared the talk thoroughly—it is not after all your talk. Reading from the screen and then commenting upon the comment, either as yourself or as you imagine the speaker might have commented, will impress.

## 6.8 Rehearsal

Rehearsing your presentation is an essential part of the preparation for a major talk. It can done at first by yourself speaking aloud, followed by a rehearsal in front of some friends or colleagues able to give you feedback. A rehearsal is particularly useful if your presentation is part of a coordinated series of talks (for example, in a proposal talk), or if the occasion is especially important (for example, a job interview or the presentation of a controversial new result for the first time).

There are several aims of the rehearsal.

- A first aim is to make sure that you are familiar with your own talk! This may seem like a trivial point, but it can happen (surprisingly often) that you notice something when the slide is projected on the big screen that you did not notice on the laptop; this can be disconcerting, to say the least. If it is a long talk with many slides, you need to know instinctively the order in which the slides are presented; it will undermine your authority if you frequently give the impression that you were not expecting *this* slide at *this* point in the talk. The rehearsal also allows you to become familiar with 'speaking the words'. This is especially important if you are nervous, or delivering the talk in another language, or if there are new, and difficult, concepts or words that you have not had to explain or use before. The act of rehearsal, like any training, prepares the brain and the other organs, so that when it comes to the crucial event, they work together in harmony to deliver a fluent performance.

- A second aim of the rehearsal is to ensure that timing is right. For some talks, it does not matter too much if the talk is 5 minutes short, or overruns by a few minutes. However, on some occasions (formal occasions, major conference talks), it is very important that the talk occupies precisely the time allotted. On such occasions, it is better to have too few slides than too many, but also to prepare a few 'filler' remarks for the last two of three slides which can be omitted if the talk takes longer than you had planned, or expanded to fill the time if you galloped through the first part.

- A third aim, for particularly important occasions, is to check with colleagues that the slides are clear, that the presentation is well-structured, that there are no obvious gaps in the argument or missing information, that there are no 'hostages to fortune' and that the conclusions are clear and comprehensive.

One of the problems with rehearsal is that it is *never* the same as the real thing. You know that this is just a practice, and the adrenaline just does not flow as freely. With practice, however, you should be able to correct for the rehearsal effect.

If possible, try to have at least one rehearsal with two or three colleagues present. The rehearsal format should be such that you have the opportunity to go through your talk without interruption; someone in the audience should act as the session chairperson, keeping track of the time. Once you have finished the talk, you should ask for general comments and then go through the presentation slide by slide. Take the comments from the audience very seriously—write them down! Try not to be offended; after all they are making the effort to help by spending their time helping you rehearse. You will have already spent several hours, perhaps days, preparing the talk, and you are too close to it—you see what you intended not what you have actually written. Your colleagues come to the talk fresh, even if they are familiar with the topic. They can see any omissions or obscurities that, once pointed out, will also be obvious to you.

So rehearse; rehearse in front of the mirror at home; rehearse in an empty lecture theatre, or with a few good friends; rehearse in your head as you travel to and from work; rehearse, if possible, with a video-camera. Even better, have a video taken of an actual presentation and review your performance critically—what seemed to work well, what did not work well, which slides were clear and which were confusing, did you look confident and convincing, were your movements natural, were the asides clearly separated from the main thrust of the presentation, did the jokes work well?

## 6.9   Dealing with questions

In regular group meetings and many small seminars, questions are asked during the course of the presentation, and the talk becomes more of an illustrated conversation. You are among friends or colleagues, and the questioning is likely to be positive and helpful, even if occasionally robust. This is just part of your normal working environment.

It is usual after a seminar, conference presentation or public lecture to allow the audience to ask questions. Indeed, one of the measures of a good talk is the level of interest that it provokes, judged by the number (and quality) of the questions. A really stimulating talk can be followed by a lengthy question and answer session, which may continue long after the final vote of thanks.

In our experience, most questions are either friendly or genuinely curious, and can be answered in a straightforward manner. The biggest difficulty can be actually hearing the question, especially in a large hall.

In a large hall it is useful to repeat the question for the benefit of the audience, and also to check that you have correctly understood it. In a smaller context, repeat the question, as if just mulling it over. We also think that it is helpful to comment on the question briefly before answering it. In one book we read on this subject, it advised *against* starting your answer with something like 'that is a very good question'. However, we take the view that if the question is interesting, even if not original, it is good to say so. We certainly would not recommend saying something like 'that is a completely trivial question, which does not deserve an answer'. If you can, explain *why* the question is a good or interesting one. This builds up a dialogue with the audience, and encourages them to ask further questions—you are doing your best to answer *their* questions.

*An example.* Suppose that you have just given a stimulating, perhaps even challenging, lecture on Time Reversal Non-Invariance.[3] An alert member of the audience might ask 'Does Time Reversal Non-Invariance mean that time travel is impossible?' Now, this is an interesting and non-trivial question, so it does no harm to acknowledge this. Your first task is to explain to the audience, and to the questioner, the subtle difference between time travel (in effect, changing $t$ to $t + (\text{or } -) \delta t$) and Time Reversal (changing $M(x,t)$ to $M^*(x,-t)$). These are two quite different operations, and it is difficult to infer much about the one from considerations about the other. You might then wish to speculate about whether time travel is theoretically possible, firstly at the micro-level (could there be quantum systems where discrete shifts in time were possible) and then at the macro-level (time travel in the science fiction sense). This may well stimulate further questions, which would allow you to discuss the nature of time, the concepts of simultaneity and relativity and the relationship between science fact and science fiction.

While the questions are often very agreeable and lead to a stimulating and open-ended debate, there are a few people who can be unsettling or disruptive, and it is necessary to develop techniques to deal with them. Here are a few common situations where the question and the questioner require some skilled handling.

---

[3]For those of you not familiar with this physics, the Time Reversal quantum mechanical operator reverses the direction of time *and* changes the matrix element to its complex conjugate, which leads to observable consequences if the matrix element has an imaginary component; there are physical processes, like the decays of neutral K-mesons, that are not invariant under Time Reversal.

*The obsessive.*   There are some people who develop an obsession with a particular theory, and who see a huge conspiracy in the scientific estab-lishment (i.e. *you*) to suppress their ideas. These are fairly easy to spot, since the question will very likely be quite long, will refer to work that they have published, or failed to have published, and promoted with almost messianic fervour. These can be on almost any subject, but 'Einstein was wrong' is a fairly common theme. Unlike the genuinely interesting ques-tion, it is not a good idea to suggest that the question is a good one, but to try to close the subject as quickly and politely as possible. For example, you might respond by saying that you have presented the conventional scientific view for which there is a great deal of experimental evidence, and leave it at that. It is probably not worthwhile trying to convince them that their position is wrong—they are obsessed with their heretical point of view, and see themselves as sadly traduced. If they persist, it is really up to the session chairperson to try to move on. You can always acknowledge their right to hold alternative views, but that you prefer to be guided by the empirical evidence and Occam's razor.

*The enthusiast.*   The enthusiast often asks quite long questions, and per-haps several of them. He or she knows quite a lot about the subject, and may well bring in information that is outside the scope of your talk or which you have chosen to leave out for a good reason. Here your approach is to acknowledge their enthusiasm and to try to answer their points as briefly as possible. It is quite likely that they are with friends, and so you should be rather careful in the way you correct any errors in their as-sumptions. Rather than pointing out the error, it is better to state clearly and succinctly the accepted view. Again, the session chairperson should take responsibility for giving other people a chance to ask questions.

*The religious.*   While the question of where God fits in is unlikely to be raised in a professional context (a departmental seminar or conference presentation), it is often raised in more general talks to the public or to schools. This is particularly likely if the subject of your talk addresses certain areas, like the Big Bang, or involves ethical issues affecting life. Unless you want to engage in a theological debate (and if you are lecturing in a theological college, it may be expected of you) it is probably advisable to answer briefly in some rather general way. Perhaps the easiest response is to effectively rule the question out of order, by saying something like 'Religious belief is a personal commitment, and it is for each individual to decide how to relate their religious beliefs to science'.

*The confused.*   Quite often, in public lectures, there will be a question that is so confused that it is difficult to find a sensible reply. There are essentially two techniques. If you can make a sensible question out of the confusion, then say something like 'I think that issue raised by this question is ...' and answer that question instead. It is possible that the questioner may wish to repeat his question, but more often they are likely to be grateful that they managed to elicit a sensible answer. If you cannot make any sense of the question, or if they come back with a repeat of their original question, it is perhaps better to say something like 'this is a rather technical question and it would probably be better to discuss it after the talk'. You will of course have to make good on your promise, but you can try to extract from the questioner the nub of their question, and away from the glare of the public platform try to provide some sort of answer or rebuttal.

*The aggressive.*   If your topic is controversial, or if your new data have just blown a huge hole in a well-known theory, you might anticipate some aggressive questioning, for example casting doubt on your credentials as a scientist or on the validity of your results. There are essentially two ways to deal with these, both based on fairly well-known quotations. The first acknowledges that the questioner has a right to hold the view that they do, but you happen to differ, perhaps quoting Voltaire 'I disagree with what you say, but uphold your right to say it'. The second asserts your position strongly, where the appropriate quote comes from John Maynard Keynes 'when the facts change, I change my mind'. Neither of these is likely to mollify the questioner, but then nothing much will. However, the audience will very likely be on your side—you have stated clearly your position with some firmness and erudition in a way that does not imply that the questioner is a complete idiot (whatever you might think). A competent chairman might allow one further comment from the questioner, but should then say something like 'well I don't think that we will agree on this one, so we had better move on to the next question ...' Of course, they may have a genuine point—'did you test for contamination with Trickisubstancin which everyone knows can mimic Realeffectomysin?' Depending upon the circumstances, you may have to concede a point, indicate that this is still work-in-progress and that more information will be available soon. Finally, you may have to rely on 'history will be my judge'; if you are feeling particularly confident, you can quote the title of Fidel Castro's book containing his defence speech at his trial in Baptista's Cuba in 1956 'History will absolve me'. However, this may not calm things down.

## 6.10   Final remarks

In this chapter, we have looked at some of the theatricality of a scientific presentation. We repeat a remark that we made earlier—style cannot replace substance. However, a well-rehearsed presentation enables the substance of the talk to be transmitted to the audience efficiently and effectively, so that while the audience thinks that is being entertained it is actually being educated.

There are many ways to develop your performance skills. You could join your local amateur dramatic society or choir, or volunteer to give talks to schools, or participate in science festivals or science fairs, or (if you are brave) try stand-up comedy in your local club on Saturday evening.

Finally, learn how to look relaxed without relaxing.

# 7
# Concluding remarks

*Life is the art of drawing sufficient conclusions from insufficient premises*
*Samuel Butler (1835–1902), Notebooks*

'Giving a talk', to colleagues, to peers or to the general public, is an important part of our professional duties as scientists. Whatever the context, we have a duty to make sure that the science is communicated clearly, and that the audience is able to understand the science at an appropriate level.

Communication of science and about science is becoming more important, for many reasons. Science is becoming more expensive—the simple (that is, cheap) experiments have been done. Science, and the technology that results from it, has brought great benefits to society, but science has also given society cause for concern. Science should be dispassionate, the results independent of cultural background and beliefs of the individual scientist; but science can also be controversial, especially when it challenges generally accepted beliefs or attitudes.

We believe that there are no absolute rules governing what makes a good slide, or how to assemble a series of good slides into a good talk. Even if there were such a recipe, how a talk is received depends upon many other things—the subject, the speaker, the venue, the size (and the mood) of the audience. There is never enough time to discuss all the details, and so we always have to make choices about what to include and what to omit. How we make these choices depends upon many things— what might be highly appropriate in one context could be completely inappropriate in another.

Nevertheless, we can think of a few principles that should help you, which have underpinned the approach taken in this book.

- Understand the scope of your talk—where do you start, what is the key point, how will you conclude, what is the message?

- Understand your hosts—why did they invite you to give the talk or, if you invited yourself, why did they agree?
- Understand your audience—why have they come to hear you, what do they know already, what do they expect to learn and what do they *need* to learn?
- Be professional—understand and be master of the technology of the presentation.
- Ensure that your slides can be seen from the back of the hall, and that you can be heard at the back of the hall.
- Be selective and be clear.
- Rehearse, review, revise and rehearse.

As you develop your skills, develop also your style. But also remember the advice that Polonius gave to his son Laertes:

> *This above all: to thine own self be true,*
> *And it must follow, as the night the day,*
> *Thou cans't not be false to any man.*
>
> William Shakespeare, *Hamlet*

Your style needs to reflect *your* personality. You need to learn how to use your natural personality to create the best effect, building upon your strengths and compensating for your weaknesses.

We have emphasized the need for preparation—we think that it takes at least ten times the duration to prepare a new talk if the topic is reasonably familiar, and much more if it is completely new. But we hope that we have also recognized the reward that follows a good talk—'thank you very much for that talk; I really enjoyed it and I learnt a lot'.

As your reputation for giving a good talk spreads, so will the invitations to give a talk will increase, and the audiences will come to expect a good talk. Preparing the talk takes time and effort, but you will gain insight into the subject and will develop a network of contacts.

# Appendix A
# Presenting complicated equations—a worked example

As we discussed in section 5.6.2, presenting complicated formulae is a challenge, even in the professional context. Of course, if nearly everyone in the audience is as familiar with the formula as you are, you can just show the equation and discuss the *particular* point that you wish to discuss—perhaps a new interpretation of its elements. However, if it is unfamiliar, you will need to take your time.

As an example, look at the equation shown in Fig. 5.38. This is clearly a complicated formula, and needs some explanation. (If you are not interested in the physics background to this equation, skip to the next paragraph.) This is the full three-flavour neutrino oscillation formula including CP-violation and matter effects. Briefly, there are three different types of neutrino, one associated with the electron, one with the muon and one with the tau—the three 'flavours' of neutrino ($\nu_e$, $\nu_\mu$ and $\nu_\tau$). However, these 'flavour-eigenstates' do not have a unique mass—they are related to three 'mass-eigenstates' ($\nu_1$, $\nu_2$ and $\nu_3$) through a mixing matrix $|\nu_\alpha\rangle = \sum_{i=1,3} U_{\alpha i}|\nu_i\rangle$. As a consequence, a neutrino created (for example, in the nuclear reactions that power the sun) as an electron neutrino can, when it is detected sometime later (say here on earth), be any of the three flavours—we say that the neutrino has 'oscillated'. There is now very strong evidence that neutrinos oscillate in this way. One of the consequences is that at most one of the neutrinos can be massless, and very likely none of them are. This has implications, among other things, for the evolution of the early universe. The phenomenology of neutrino oscillations is very rich, which is another way of saying that the equations are rather forbidding at first sight.

Unless you are giving the summary talk at ICOWENOF (the International Conference of World Experts on Neutrino Oscillation Formulae), where the audience will have seen the full oscillation equation in every one of the previous fifty talks, you may be fairly certain that some in the

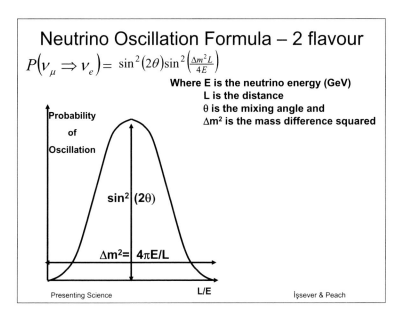

**Fig. A.1** The 2-flavour neutrino oscillation formula—the first step in developing the full 3-flavour formula shown in Fig. A.4.

audience will find it a little daunting. Talking them through it might take quite a long time.

An alternative is to break the formula down into a series of four slides. In the first slide (see Fig. A.1), you present the *two*-flavour oscillation formula, together with a graph showing the *two* parameters that describe the effect. You can explain that the *mixing angle* $\theta$ governs the *amplitude* of the oscillation from one flavour (say muon neutrino) to another (say electron neutrino), and that the *mass difference squared* $\Delta m^2$ governs the oscillation frequency as a function of the *length* $L$ that the neutrino of *energy* $E$ has travelled since it was created. This is relatively simple and easy to understand.

Next, you show Fig. A.2, which is considerably more complicated, and represents the three-flavour oscillation amplitude, this time as a function of *three* mixing angles $\theta_{12}$, $\theta_{13}$ and $\theta_{23}$, and *three* mass differences squared $\Delta m_{21}^2$, $\Delta m_{31}^2$ and $\Delta m_{32}^2$. You can make several simple remarks.

1. Although the equation is considerably more complicated than the previous one, it shares many common features.

2. To reduce the amount of space occupied, you have had to introduce a more compact notation—$s_{ij} = \sin \theta_{ij}$ and $c_{ij} = \cos \theta_{ij}$

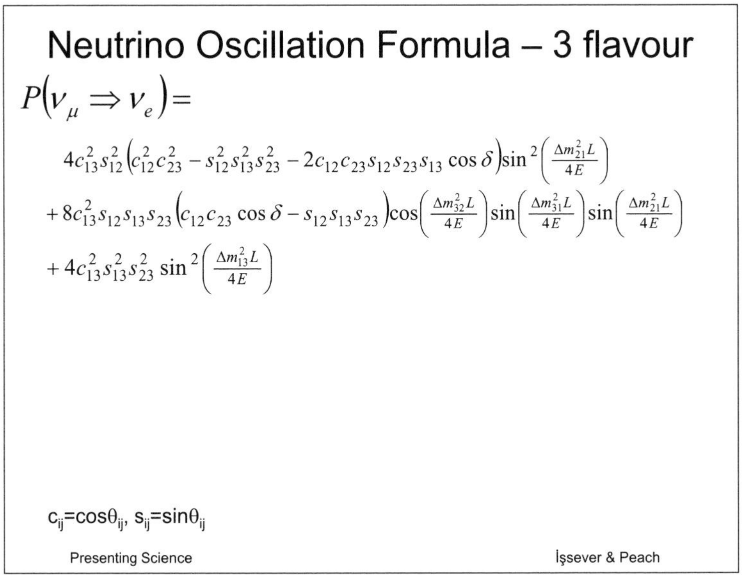

---

### Neutrino Oscillation Formula – 3 flavour

$$P\left(\nu_\mu \Rightarrow \nu_e\right)=$$

$$4c_{13}^2 s_{12}^2 \left(c_{12}^2 c_{23}^2 - s_{12}^2 s_{13}^2 s_{23}^2 - 2c_{12}c_{23}s_{12}s_{23}s_{13}\cos\delta\right)\sin^2\left(\frac{\Delta m_{21}^2 L}{4E}\right)$$

$$+\, 8c_{13}^2 s_{12}s_{13}s_{23}\left(c_{12}c_{23}\cos\delta - s_{12}s_{13}s_{23}\right)\cos\left(\frac{\Delta m_{32}^2 L}{4E}\right)\sin\left(\frac{\Delta m_{31}^2 L}{4E}\right)\sin\left(\frac{\Delta m_{21}^2 L}{4E}\right)$$

$$+\, 4c_{13}^2 s_{13}^2 s_{23}^2 \sin^2\left(\frac{\Delta m_{13}^2 L}{4E}\right)$$

$c_{ij}=\cos\theta_{ij},\ s_{ij}=\sin\theta_{ij}$

Presenting Science                                   İşsever & Peach

---

**Fig. A.2** The 3-flavour neutrino oscillation formula, without CP-violation and matter effects—the second step in developing the full 3-flavour formula shown in Fig. A.4.

3. The first and third terms are just like the simple two-flavour oscillation formula, except that amplitude is now a complicated function of the three mixing angles $\theta_{12}$, $\theta_{13}$ and $\theta_{23}$, rather than a single angle. However, the amplitude is a constant. The oscillation frequencies are governed by two different mass-differences squared, mixing the first and second and first and third generations respectively.

4. The second term is more complicated, and represents the *interference* between the other two oscillation frequencies. Again, the amplitude is a constant, depending only upon the angles. The oscillation term has a richer structure, and has been written here as a function of three mass differences squared, to emphasize the origin of the mixing, but only two of the mass differences are independent, since $\Delta m_{32}^2 = \Delta m_{31}^2 - \Delta m_{21}^2$.

5. In the absence of CP-violation, $\cos\delta = 1$. In the presence of CP-violation ($\delta \neq n\pi, n = 0, 1, \ldots$), it is still possible *in principle* to observe CP-violating effects by measuring only CP-conserving processes such as $\nu_\mu \to \nu_e$ and $\nu_\mu \to \nu_\tau$ oscillations for both neutrinos and antineutrinos, but that this is not practicable because it requires un-

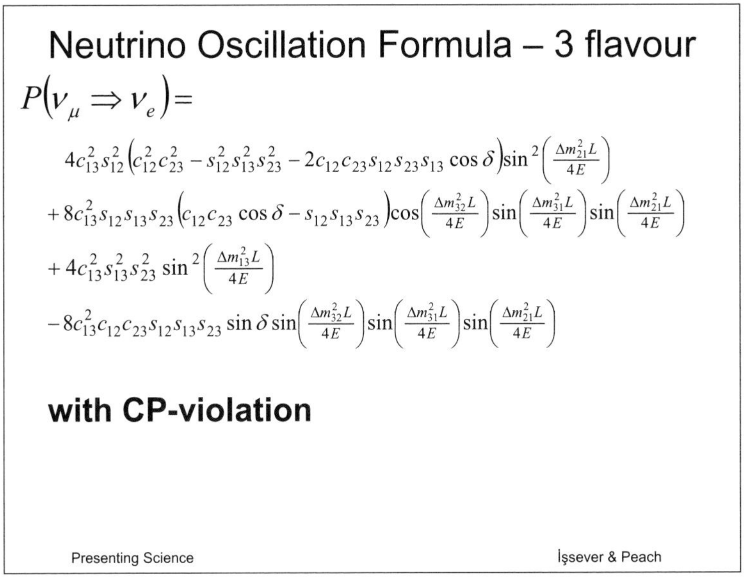

## Neutrino Oscillation Formula – 3 flavour

$$P\left(\nu_\mu \Rightarrow \nu_e\right)=$$

$$4c_{13}^2 s_{12}^2 \left(c_{12}^2 c_{23}^2 - s_{12}^2 s_{13}^2 s_{23}^2 - 2c_{12}c_{23}s_{12}s_{23}s_{13}\cos\delta\right)\sin^2\left(\frac{\Delta m_{21}^2 L}{4E}\right)$$

$$+8c_{13}^2 s_{12}s_{13}s_{23}\left(c_{12}c_{23}\cos\delta - s_{12}s_{13}s_{23}\right)\cos\left(\frac{\Delta m_{32}^2 L}{4E}\right)\sin\left(\frac{\Delta m_{31}^2 L}{4E}\right)\sin\left(\frac{\Delta m_{21}^2 L}{4E}\right)$$

$$+4c_{13}^2 s_{13}^2 s_{23}^2 \sin^2\left(\frac{\Delta m_{13}^2 L}{4E}\right)$$

$$-8c_{13}^2 c_{12}c_{23}s_{12}s_{13}s_{23}\sin\delta\sin\left(\frac{\Delta m_{32}^2 L}{4E}\right)\sin\left(\frac{\Delta m_{31}^2 L}{4E}\right)\sin\left(\frac{\Delta m_{21}^2 L}{4E}\right)$$

### with CP-violation

Presenting Science         İşsever & Peach

**Fig. A.3** The 3-flavour neutrino oscillation formula, with CP-violation but without matter effects—the third step in developing the full 3-flavour formula shown in Fig. A.4.

achievable precision on the measurement of the three mixing angles $\theta_{12}$, $\theta_{13}$ and $\theta_{23}$.

Then, you introduce Fig. A.3, which is the same as Fig. A.2, except that you have added a term in $\sin\delta$, which is CP-violating. There are essentially four remarks that you need to make on this development.

1. There is a clear similarity between this new term and the second term, which is not surprising since this comes from the imaginary part of the interference amplitude.
2. When changing either from neutrinos to antineutrinos, the sign of this term changes.
3. When comparing $\nu_\mu \rightarrow \nu_e$ with $\nu_e \rightarrow \nu_\mu$, the sign of this term changes.
4. If *any* of the angles $\theta_{12}$, $\theta_{13}$ or $\theta_{23}$ is equal to $n\pi/2, n = 0, 1, \ldots$, there is no CP-violation.

Finally, you introduce the full equation (see Fig. A.4), including the matter effects. Again, there are a few remarks to make.

1. Adding matter effects of the earth modifies the third term and introduces a fifth term.

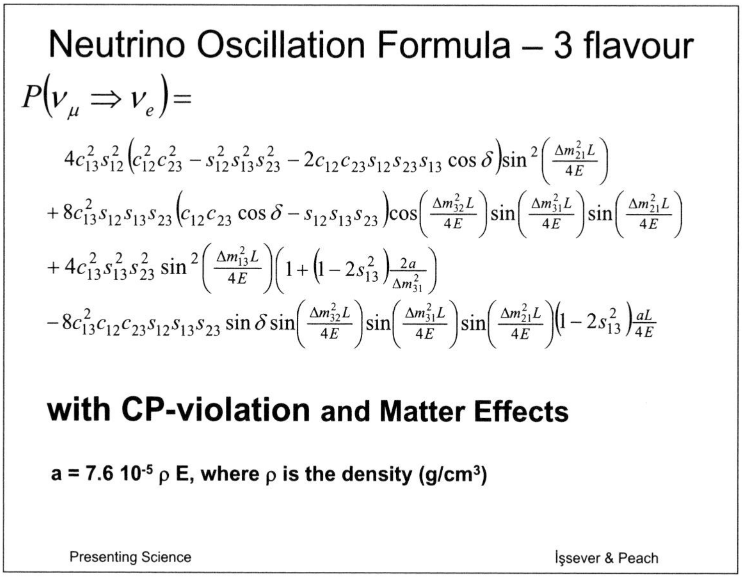

---

## Neutrino Oscillation Formula – 3 flavour

$$P\left(\nu_\mu \Rightarrow \nu_e\right)=$$

$$4c_{13}^2 s_{12}^2 \left(c_{12}^2 c_{23}^2 - s_{12}^2 s_{13}^2 s_{23}^2 - 2c_{12}c_{23}s_{12}s_{23}s_{13}\cos\delta\right)\sin^2\left(\frac{\Delta m_{21}^2 L}{4E}\right)$$

$$+ 8c_{13}^2 s_{12}s_{13}s_{23}\left(c_{12}c_{23}\cos\delta - s_{12}s_{13}s_{23}\right)\cos\left(\frac{\Delta m_{32}^2 L}{4E}\right)\sin\left(\frac{\Delta m_{31}^2 L}{4E}\right)\sin\left(\frac{\Delta m_{21}^2 L}{4E}\right)$$

$$+ 4c_{13}^2 s_{13}^2 s_{23}^2 \sin^2\left(\frac{\Delta m_{13}^2 L}{4E}\right)\left(1 + \left(1 - 2s_{13}^2\right)\frac{2a}{\Delta m_{31}^2}\right)$$

$$- 8c_{13}^2 c_{12}c_{23}s_{12}s_{13}s_{23}\sin\delta\sin\left(\frac{\Delta m_{32}^2 L}{4E}\right)\sin\left(\frac{\Delta m_{31}^2 L}{4E}\right)\sin\left(\frac{\Delta m_{21}^2 L}{4E}\right)\left(1 - 2s_{13}^2\right)\frac{aL}{4E}$$

### with CP-violation and Matter Effects

**a = 7.6 10⁻⁵ ρ E, where ρ is the density (g/cm³)**

Presenting Science    İşsever & Peach

---

**Fig. A.4** The full 3-flavour neutrino oscillation formula, including CP-violation and matter effects—the final step.

2. The effect is only significant at large distances.
3. The sign of the effect changes in going from neutrinos to antineutrinos.
4. Because of the modification to the third term, this can be used to measure the sign of $\Delta m_{31}^2$.

While this looks to be a long way of introducing the formula, in fact it can be completed quite quickly for the professional audience. The point is that, at each step, you have explained carefully how it relates to the previous step, which they understood, and that each step introduces some additional complexity but also additional, and rich, phenomenology. Those who need to will work through the complexity later, once they have understood the phenomenology.

An alternative way of presenting this equation to a professional audience, which has some advantages, is shown in Figs A.5 and A.6. Here, the first slide can be used to show the general structure of the equation. The first and third terms, which are multiplied by the generic functions $A(\theta_{12}, \theta_{23}, \theta_{13}, \delta)$ and $C(\theta_{12}, \theta_{23}, \theta_{13}, \delta)$, describe the simple oscillation terms. The second and fourth terms, multiplied by $B(\theta_{12}, \theta_{23}, \theta_{13}, \delta)$ and $D(\theta_{12}, \theta_{23}, \theta_{13}, \delta)$, describe the real and (CP-violating) interference

## Neutrino Oscillation Formula – 3 flavour

$$P(\nu_\mu \Rightarrow \nu_e) = A(\theta_{12},\theta_{23}\theta_{13},\delta)\sin^2\left(\frac{\Delta m_{21}^2 L}{4E}\right)$$

$$+ B(\theta_{12},\theta_{23}\theta_{13},\delta)\cos\left(\frac{\Delta m_{32}^2 L}{4E}\right)\sin\left(\frac{\Delta m_{31}^2 L}{4E}\right)\sin\left(\frac{\Delta m_{21}^2 L}{4E}\right)$$

$$+ C(\theta_{12},\theta_{23}\theta_{13},\delta)\sin^2\left(\frac{\Delta m_{13}^2 L}{4E}\right)$$

$$+ D(\theta_{12},\theta_{23}\theta_{13},\delta)\cos\left(\frac{\Delta m_{32}^2 L}{4E}\right)\sin\left(\frac{\Delta m_{31}^2 L}{4E}\right)\sin\left(\frac{\Delta m_{21}^2 L}{4E}\right)\left(1+\left(1-2s_{13}^2\right)\frac{2a}{\Delta m_{31}^2}\right)$$

$$+ E(\theta_{12},\theta_{23}\theta_{13},\delta)\cos\left(\frac{\Delta m_{32}^2 L}{4E}\right)\sin\left(\frac{\Delta m_{31}^2 L}{4E}\right)\sin\left(\frac{\Delta m_{21}^2 L}{4E}\right)\left(1-2s_{13}^2\right)\frac{aL}{4E}$$

### with CP-violation and Matter Effects

**a = 7.6 10⁻⁵ ρ E, where ρ is the density (g/cm³)**

Presenting Science    İşsever & Peach

**Fig. A.5** An alternative way of presenting the equation in Fig. A.4, where the amplitude of the oscillating terms is replaced by functions *A–E*.

## Amplitude formulae

$$A(\theta_{12},\theta_{23},\theta_{13},\delta) = +4c_{13}^2 s_{12}^2\left(c_{12}^2 c_{23}^2 - s_{12}^2 s_{13}^2 s_{23}^2 - 2c_{12}c_{23}s_{12}s_{23}s_{13}\cos\delta\right)$$

$$B(\theta_{12},\theta_{23},\theta_{13},\delta) = +8c_{13}^2 s_{12}s_{13}s_{23}\left(c_{12}c_{23}\cos\delta - s_{12}s_{13}s_{23}\right)$$

$$C(\theta_{12},\theta_{23},\theta_{13},\delta) = +4c_{13}^2 s_{13}^2 s_{23}^2$$

$$D(\theta_{12},\theta_{23},\theta_{13},\delta) = -8c_{13}^2 c_{12}c_{23}s_{12}s_{13}s_{23}\sin\delta$$

$$E(\theta_{12},\theta_{23},\theta_{13},\delta) = -8c_{13}^2 s_{13}^2 s_{23}^2$$

$$c_{ij} = \cos(\theta_{ij})\,;\, s_{ij} = \sin(\theta_{ij})$$

Presenting Science    İşsever & Peach

**Fig. A.6** The amplitude functions associated with Fig. A.5

terms. The last term, multiplied $E(\theta_{12}, \theta_{23}, \theta_{13}, \delta)$, describes the effect of the passage through the earth, which also modifies the fourth term $(1 + (1 - 2s_{13}^2)a/\Delta M_{31}^2)$. The details of the amplitude functions can then be briefly discussed using Fig. A.6.

# Appendix B
# Some PowerPoint tips

This is not a comprehensive manual for using PowerPoint, which is a very powerful presentation manager, with very many advanced features, and some weaknesses. We have collected here a few things that we have found useful in giving talks, and in preparing this book. In fact, one of the benefits to us of preparing this book is that we have been able to share some of these with each other.

## B.1   Blanking out the screen

Sometimes, it is useful to display a blank screen—you may wish to surprise your audience with the title slide, for example, or you may wish temporarily to divert their eyes from the screen while you show a demonstration. One way of doing this, of course, is to include a blank or neutral frame. Sometimes, we see people disconnecting the cable to the data projector or disabling the display on the laptop to achieve this effect. In fact, PowerPoint has an in-built function that does *exactly* what you want, and drives the data projector with a blank screen. Type 'B' on the keyboard, and the screen goes blank.[1] Type 'B' again, and the screen comes back. As Michael Caine used to say, 'not many people know that'.

## B.2   Running out of time

This is where PowerPoint and other presentation managers are less good than old fashioned plastic foils. With foils, it is relatively easy to skip a few and get to the next important point, but with the presentation managers it is awkward. If you know your presentation by heart, there are two tricks with PowerPoint that can help. Firstly, if you type a number when in display mode, followed by 'Enter', PowerPoint will go straight to that slide. Secondly, if you type in '+n', PowerPoint will skip n-1 (!) slides—i.e. to go to the next slide, enter '+2'. Note that entering '-n' does

---

[1] This only works if you are in full-screen mode.

not achieve a skip backwards, but exits the presentation. If you do not have a photographic memory, it is useful to print out a version of your presentation with sixteen or twenty-five frames per page, so that you can quickly skip to slide n if needed.

## B.3   Saving space with pictures

Photographs and figures can take up a lot of space. The size of the file can be reduced dramatically by compressing pictures and plots. You can do this with the compress button in the picture tool bar in PowerPoint. (Note that you should keep a separate full-resolution copy of the picture elsewhere if you intend to use it in a publication.)

## B.4   PowerPoint and equations

In this book, we have used PowerPoint as a presentation manager, because it is widely available. For most purposes, it is more than adequate—simple displays of text and graphics, with modest animation capabilities. However, for equations it has shortcomings. Anything more complicated than $F = ma$ is difficult, if it involves both superscripts and subscripts, overlining, differential operators, etc. Even if it involves just superscripts *or* subscripts, line spacing becomes a problem.

Microsoft Equation 3.0 (INSERT/OBJECT) solves some of these problems, but lacks flexibility over font size, colour, emphasis and animation. MathType has some more functionality (the ability to have different colours for example) but it is still difficult to embed equations neatly into text.

It is possible to find software like TEXPoint that allows you to embed TEX or LATEX typeset equations into PowerPoint, but this also has limitations.

It is always possible to create the equation that you want in another package, and either include as a screen grab (using the PrtSc button), or by converting to a standard graphics format like JPEG and including as a picture, but it is all really rather tedious. If you really have a lot of equations, you might consider abandoning PowerPoint altogether and using a different presentation manager (LATEX for example), but this can bring other problems.

# Appendix C
# Meeting the media

Presenting science to and through the media (press, radio, television) is very important, and is becoming increasingly so. However, we have to understand that the media imperative is very different from that of the scientist's normal working environment. Failure to understand and accommodate this difference can lead to personal embarrassment, professional discomfort and public humiliation.[1] So, if you have ambitions or expect to be involved with the media, our first piece of advice is to enroll as soon as possible in a **media training** course specifically aimed at scientists. In this short book, we can only offer some rather general guidance.

When dealing with journalists, it is vital to remember that it is the *journalist* who is your audience and not the general public; the general public is the *journalist's* audience. Of course, you hope to get *your* message through the journalist to the general public, but it will be the *journalist's* version of your message. If you are skilful in your communication with the journalist, the message that the general public receive will be one which you recognize and endorse. If this is *not* the message that you wanted conveyed, then sadly it is probably *your* fault—you failed to understand the media context.

The most important thing to understand and remember is that the journalist wants a good story. It needs a good *hook*, one that will interest the readers, listeners or viewers and impress the editors. On the other hand, the journalist does not want to make mistakes or to misrepresent the truth, but equally does not want to *invent* anything—the story will be based on what *you* said. So again, if the story is not the one that you wanted, the most likely explanation is that *you* failed to make your message clear.

There are, roughly speaking, two different scenarios. In the first, you

---

[1] Although not really in the scientific domain, a good example of what can go wrong if this difference is not fully understood is provided by the story of the forged 'Hitler Diaries' published by the German magazine *Stern* in April 1983.

have issued a press release or published an interesting paper, and this has excited some press interest; you initiated this, and you must prepare your *presentation* in exactly the same way, and with the same attention to detail, as you would a seminar. In the second, the press have contacted you out of the blue; you cannot prepare in advance, but you can still 'play for time'. These two scenarios are discussed below.

## C.1   After a press release

Many press releases about scientific results fail to make an impact in the media because of a very simple mistake—the press release addresses the *scientific* context and not the *journalistic* context. Quite often (especially in the local press, but occasionally in the national press), the published story is little more than a lightly edited version of the press release. If this is what the journalist intends to do, it is unlikely that they will make contact—they have all the information they need, and you (or your press officer) have done a good job.

If you (or your press officer) have done an *excellent* job, a journalist may contact you for an interview as part of a longer piece on your work. You must prepare for this *before* you issue the press release. Journalists work to tight deadlines (always ask what the deadline is, and don't be surprised if the reply is 'now') so there will not be time to prepare after the contact. Try to anticipate the questions, and prepare answers. One tip is to prepare a set of 'slides' (in your favourite presentation manager) each with a possible question and an appropriate answer. The point is not to read from these 'cheat sheets', but the act of preparing them means that you have partially formed answers ready in your mind. While the actual words that you use will be different, the basic thought will be close to what you intended.

If you are lucky, you might have a face-to-face interview with the journalist, but very often it will be over the telephone. Even more terrifying is the radio interview live 'down the line'—all you have is a microphone, a set of earphones and a blank wall. One difficulty is that, although you have potentially a large audience, you are actually alone. It is therefore vital that you think clearly about your key messages, and that you rehearse a way of getting that message over economically—in the famous 'sound bites'. (Even if it is just a journalist on the phone, he or she will very likely be writing your words down in shorthand; leanness of style will be appreciated.)

Although (in our experience) the journalist will wish to present a positive story ('breakthrough in the fight against cancer' or 'unlocking the secrets of the universe'), the journalistic imperative is likely to prompt a few questions which present some dangers.

'Isn't this rather expensive?'

'What would happen if the experiment went wrong—could there be an explosion?'

'Are there any risks in this procedure?'

You need to be prepared for these questions—a careless answer could undermine your message; prepare your response.

Finally, remember that if you provide journalists with good stories, you are doing them a favour ... but don't let them know that you know this!

## C.2  The unsolicited contact

There are many reasons why you might be contacted by a journalist for a comment on a current scientific controversy. The difference between this and the previous situation is that you have not had time to think through your response. It is tempting to treat this encounter as if it was a discussion with a colleague, but this would be wrong—the journalist is looking for an interesting story, and there is always more interest in conflict than in accord. So, be careful!

Even though the journalist's deadline might be quite short, it is usually better to play for time, just so that you can organize your thoughts. A useful tactic is to insist that the interview is 'off the record', but that you will provide a quote 'on the record' if that is useful. This reassures the journalist that he or she is not wasting their time; at the end of the day, they like to write something along the lines of 'Professor X said that this was the breakthrough of the decade'. Note that giving an interview 'off the record' does not mean that your views will not be made public—all that it means is that the remarks will not be attributed to you.

You can use the conversation 'off the record' to try to find out what the journalist's angle is, and during this you may be able to devise a quote that is suitable, or the journalist might suggest a quote from the 'off-the-record' discussion that could be attributed to you. Alternatively, you might ask for a few minutes to review the paper, and return the call—if you do this, you *must* return the call, and provide a quote. (Of course, the journalist might not use it, but that is not your problem.)

If you can, it is good if you can check the words that are attributed to you, and the context. Some journalists are forbidden by the house rules of their journal from consulting in this way, but often they are receptive to the line that you can check the copy for factual accuracy. (As noted above, journalists do not like to make mistakes, and few have sufficient scientific knowledge to be expert in everything.) If you *are* provided the opportunity to 'check the facts' it is very important that you restrict yourself to ensuring that the article is scientifically correct, and that you do *not* try to change the 'slant' of the story—even if you disagree with some of it—it is after all the *journalist's* story.

## C.3   A final word on the media

Dealing with the media poses special challenges, but can also bring significant rewards. An article in a national newspaper or a 10-minute slot on BBC's Radio 4 can reach an audience of hundreds of thousands or millions—how many lectures would you have to give in half-empty halls to reach that number of people? As with the public lecture, you can generally expect a fair hearing, unless you are involved in one of the major controversies. And as with the public lecture, it is important that you are properly and appropriately prepared. If you can establish a reputation for delivering what the journalists want, you can expect to be contacted quite regularly. But be careful—there can also be professional jealousy if you attract a lot of media attention.

# Index

academic lectures, 10–11, 34–5, 39–40
agendas,
    as 'prior notice,' 27
    workshop, 30
    position on, 48
anecdotes, *see humour*
animation, 6, 7, 44, 67–71
    between-slide, 68
    embedded clips, 70–71
    for emphasis, 64, 66, 85
    in-slide, 69–70
    reveal animation, 60
audibility, *see* microphones
audiences,
    alienation of, 21, 91
    assessment, 38–40
        motivation, 38–9
        knowledge, 39–40
    comments from, 97
    communication with, 1, 2
    composition, 26
    defining context, 25, 26, 28
    expectations, 5, 18, 28, 31
    international/foreign, 8, 91
    job interview seminars, 37
    knowledge level, 28, 29, 36
        assessment of, 39–40
        experts, 14, 29, 83
        and jargon, 21
        lay audiences, 4, 14, 22–3
        scientific, 22
        and structure of content, 14
    leaving, 5, 31, 32

and the media, 113, 114, 116
needs of, 2, 9, 19, 20–1, 36
of one, 32
and presenter position, 84–6
size, 4, 5, 30, 31, 41
    and audibility, 86–8
understanding your audience 4–5,
    9, 29, 103

background information, 20, 34
backgrounds *see under* slides
body language, 15–16
    *see also* position to audience
borders, 49, 54–6
boxes, 53, 58–9, 60,
    highlighting information, 62, 66–7
    resizing, 62
bullet points, 58, 59, 60, 61, 62, 63–4
    levels of bullets 19
    posters, 78
    symbols, 63

colloquia, 4, 27–9
    *see also* seminars
colour, 44–6
colour blindness, 46
compression of pictures, 112
conclusions slides, 9, 32
conferences, 30–3
    audience assessment, 14, 39
    equations in, 75–6
    panel discussions, 32
    parallel sessions, 31–2

conferences (*cont.*)
  plenary sessions, 31
  poster sessions, 32
  summaries, 30, 32–3
  size, 30
  types, 30
  *see also* telephone/video conferences
contents slide, 9, 19–20, 32
context, 25–6
  common contexts, 26–38
copyright, 18, 92
corporate style, 8–9, 57–8

departmental seminars, 39
dialogue/discussion with audience, 27,
    40, 41, 77, 98
dress code, 90

equations, 71–7
  example of presentation, 104–10
  formal derivation, 72–4
  in PowerPoint, 112
essential points, 1

figures, 64–5, 112
font(s), 8, 9, 44–7
  colour, 44–6
  embedding, 7, 46–7
  fancy, 46–7
  type/style, 44, 46
foreign language, 7–8, 94, 96
formal lectures, 42
frames, 47–58
  blank, 111
  of posters, 78

general interest seminars, 22, 29, 30
  audiences, 39
grant proposal talks, 10, 33–4
graphs *see* plots
group meetings, 26–7
  equations in, 75–6
  questions in, 103
  slide style, 44, 45

handouts, 36
humour, 90–3

accidental/failed, 87, 89, 90
bonding/excluding, 32, 91
as 'punctuation', 15, 36, 90
spontaneity, dangers of, 91–2
timing, 91
visual, 92–3

interviews
  job, 37–8
  media, 114–6
introductions, 5, 13, 15, 17,19, 20–1
  colloquia/seminars, 28, 29
  combining with outline, 20
  public lectures, 36
  timing, 42

jargon, 20, 86

key result, 9, 22, 23, 33
  and timing, 41
key message, 98, 114
  on poster, 79
  on slides, 60, 61

language, 7–8, 86
  body language, 15–16
  foreign languages, 7–8, 94, 96
  jargon, 21, 86
laser pointers, 27, 60, 65, 85, 86,
    88–9
LATEX, 112
lay audiences, 4, 14, 22–3, 40
  *see also* popular/public talks
leaflets, 36
length of presentation, 5, 14, 40–1

media (presentation), 6, 7
media (radio, television, press), 113–6
  interviews, 114–6
  timing of presentations, 42
media training course, 113
microphones, 86–8
  fixed podium, 87
  hand-held, 88
  head mounted, 87
  lapel, 87
  radio, 15

motivation
  of audience, 38
  for invitation, 28
  for work, 21

nervousness, 7, 8, 42, 83, 93–5
notes, 8, 25, 94, 95

oratorical skills, 36
overhead projectors, 7, 49, 84–5

panel discussions, 32
parallel session talks, 31–2, 40
photographs, 36, 48, 92, 112
  examples, 51–2
pictures, 50–1, 52–3, 64, 65
plenary session talks, 31, 40
  summaries, 33
plots, 58, 59, 60, 63, 64
pointers, laser, 27, 60, 88–9
  for indistinct plots, 65
  and position to audience, 85, 86
popular/public talks
  equations in, 71, 76–7
  entertainment in, 90
  audiences, *see* lay audiences
  structure, 15, 18, 35–7
position to audience, 84–6
poster sessions, 32, 77
posters, 77–9
  audience, 32
  flow, 78
  handouts, 36, 79
  layout, 78–9
  pictures/plots, 78
  sizes, 77–8
  style, 80
  words, number of, 78
PowerPoint, 2
  fonts, embedded, 47
  Notes, 8
  slide layout, 47
  tips, 111–2
slide numbers, 95
preparation 4, 82
  and context, 25, 43
  laser pointer, use of, 88–9

for nervousness, 93–5
position to audience, 84–6
practising, 83–4, 95–7
and presentation media, 6
public address system, 87–8
self assessment, 82–3
time, 4
title slide, using, 17
presenter
  body language, 15–16
  position to audience, 84–6
  strengths/weaknesses, 82–3
  voice, 30, 82, 86
    modulation, 87, 88
presentation(s), 82–101
  breaks/pauses, 15
  and communication, 1
  of own work, 31
presentation management device, 89
professional seminars, 15, 71, 75–6
project proposal talk, 33
props, 36
public/popular lectures, 15, 18, 35–7,
    71, 76–7, 90
  audiences, *see* lay audiences
  questions in, 97, 99, 100

questions, 35, 36,
  dealing with difficulties, 84, 97–101,
    115
  and event structure, 5, 6, 25
  media presentations, 114, 115
  preparation for, 10, 30
  and slide numbering, 48
  timing, 35, 36, 41

radio presentations, 42, 113–6
rehearsal, 33, 41, 86, 95–7
report presentation, 34
review talk, 31

schools lecture, 37, 76–7
self-awareness, 82–3
seminars, 4, 27–9, 75–6
  audiences, 39
  departmental, 39
  general interest, 22, 29, 30

seminars (*cont.*)
  job interview, 37
  professional, 15, 71, 75–6
  *see also* colloquia
slides
  animation, 6, 7, 44, 67–71
    between-slide, 69
    embedded clips, 69–71
    for emphasis, 64, 66, 85
    in-slide, 68–9
    reveal animation, 60
  backgrounds, 8, 46, 47, 48–9
    examples, 50–3, 54–5, 57
  backup, 10, 27, 28
  borders, 49, 54–6
  boxes, 53, 58–9, 60, 62, 66–7
    hand-drawn for emphasis, 89
  bullet points, 58, 59, 60, 61, 62,
    63–4
    posters, 78
    sub-bullets, 20
    symbols, 63
  colour, 44–6
  conclusions, 9, 32
  contents, 9, 19–20, 32
  equations, 71–7, 104–10
  figures, 64–5, 112
  font, 44–7
    colour, 44–6
    fancy, 46–7
    type/style, 44, 46
    number, 48
  photographs, 36, 48, 92, 112
    examples, 51–2
  pictures, 50–1, 52–3, 64, 65
  plots, 58, 59, 60, 63, 64
  structure, 13–14
  tables, 34, 66–8
staff meetings, 26–7
stagefright, 83
structure
  of event, 5
  of presentation, 13–24
    background, 21
    introduction, 17, 20–1

methodology, 22
  outline, 19–20
  results, 22–3
  slides, 13–14
  summary/outlook, 23–4
  title slides, 16–19
  transitions, 15–16
style, 28, 44–81
  consistency, 9
  developing, 26, 103
  and personality, 2, 103
  and presentation media, 6
  *see also* corporate style
summaries
  presentation summary, 23–4, 42
  conference summary, 30, 32–3
summer school lectures, 34–5, 39–40

technical terms, 20, 86
telephone/video conferences, 6, 37,
    48, 88
  media interviews, 114
television presentations, 42, 80, 113–6
TEXPoint, 112
timing, 33, 41–2
  embedded animation, 70
  humour, 91
  rehearsal, 96
title slides, 15, 16–19, 20, 55
titles
  of posters, 78
  of slides, 14, 16
  of talk, 10

unconventional beginning, 18

venue, 5
  room layout, 85
video conferences, 6, 37, 48, 88
voice, 30, 82, 86
  modulation, 87, 88

weekly progress report, 15, 17
workshops, 30–3
  audience assessment, 39